Lecture Notes in Earth Sciences 75

Editors:
S. Bhattacharji, Brooklyn
G. M. Friedman, Brooklyn and Troy
H. J. Neugebauer, Bonn
A. Seilacher, Tuebingen and Yale

T0074236

Springer-Verlag Berlin Heidelberg GmbH

Hildegard Westphal

Carbonate Platform Slopes – A Record of Changing Conditions

The Pliocene of the Bahamas

With 56 Figures and 10 Tables

 Springer

Author

Dr. Hildegard Westphal
Rosenstiel School for Marine and Atmospheric Sciences
Division of Marine Geology and Geophysics
4600 Rickenbacker Causeway, Miami, FL 33149-1098, USA
E-mail: hwestphal@rsmas.miami.edu

"For all Lecture Notes in Earth Sciences published till now please see final pages of
the book"

Cataloging-in-Publication data applied for

Die Deutsche Bibliothek - CIP-Einheitsaufnahme

Westphal, Hildegard:
Carbonate platform slopes - a record of changing conditions : the
pliocene of the Bahamas ; with 10 tables / Hildegard Westphal. -

(Lecture notes in earth sciences ; 75)
ISBN 978-3-540-64646-4   ISBN 978-3-540-69176-1 (eBook)
DOI 10.1007/978-3-540-69176-1

ISSN 0930-0317
ISBN 978-3-540-64646-4

© Springer-Verlag Berlin Heidelberg 1998
Originally published by Springer-Verlag Berlin Heidelberg New York in 1998

Typesetting: Camera ready by author
SPIN: 10680137   32/3142-543210 - Printed on acid-free paper

To Stefan who shares my love for science

Preface

This book represents the author's doctorals thesis. Therefore, although it is a monograph, it represents the results of numerous discussions and cooperations.

First of all I am grateful to Christian Dullo (GEOMAR, Kiel/Germany) for supervising my work on this thesis, for valuable suggestions, and for providing support whenever I needed it. I want to thank John J. G. Reijmer (GEOMAR, Kiel), who initiated my work on core CLINO, who was involved in the sampling of core CLINO in Miami, and who with many discussions and hints was of continuous help.

I am especially thankful to Robert N. Ginsburg and Gregor P. Eberli (Comparative Sedimentology Laboratory at Rosenstiel School of Marine and Atmospheric Sciences, University of Miami) for giving me the opportunity to study the core CLINO of the Bahamas Drilling Project, and for providing information on the work done by the Miami group. Jeroen Kenter introduced me to the core and suggested the selection of the intervals that were studied in this thesis. Flavio Anselmetti, Gregor Eberli, Don McNeill, Leslie Melim, and Peter Swart were extremely helpful in providing information, unpublished figures, and data on the Bahamas Drilling Project. To Leslie, I am particularly indebted for many discussions and for a review of the diagenesis part of the present manuscript. Peter Swart performed some stable isotope measurements for calibrating the data of the present study with the data previously obtained at the RSMAS. To Karin Bernett I am indebted for some later sampling of the core. Mike Grammer and Pamela Reid kindly provided recent fecal pellets and *Halimeda* specimens from the Bahamas and Florida for comparison with the fossil ones present in CLINO.

Christian Samtleben and Axel Munnecke (Universität Kiel) introduced me to the art of scanning electron microscopy. The numerous fruitful discussions strongly influenced the development of this thesis. To Axel, I am grateful for directing my focus on the diagenesis of micritic limestones and the development of microspar. Also, he kindly reviewed an earlier version of the present manuscript.

Bob Goldstein (University of Kansas/Lawrence) did the fluid inclusions measurements. Discussions with him on a predictive diagenetic model were a lot of fun.

The co-operation with Martin Head (Toronto University), who did the palynological examinations of my samples and took the micrographs of the palynological slides, was highly interesting. I am very grateful that he taught me a lot on palynomorphs. Thanks to Thomas Servais (Université de Lille) for rising my interest for palynomorphs. Wolfram Brenner (GEOMAR Kiel) and Hans Gocht (Universität Tübingen) helped me with the work on dinoflagellate cysts.

Klaus Vogel (Universität Frankfurt) helped with the determination of burrows in pellets. On the determination of foraminifers I have learned from Christoph Hemleben, Ulrike Wieland (both Universität Tübingen), and Dorothee Spiegler (GEOMAR Kiel). Xin Su (Beijing University) had a look at a coccolith specimen.

The measurements of stable isotopes were performed in the laboratory of Michael Joachimski at the University of Erlangen/Germany. Thanks a lot! To Tom Aigner and Markus Schauer (both Universität Tübingen) I am grateful for the possibility to test the luminescence of my samples with the cathodoluminescence device of their department.

To Adam Vecsei (Universität Freiburg) I am indebted for the review of the sediment input-part of this thesis. The entire thesis greatly benefited from peer reviews of related manuscripts by Julian E. Andrews, Roger Barnaby, Robin G. C. Bathurst, Mark Harris, Zak Lasemi, Leslie Melim, and Bruce W. Sellwood.

Many thanks are due to Werner Ricken (Universität Köln) for various discussions on carbonate diagenesis and for reviewing one of the manuscripts prior to submission.

Stefan Bornholdt (Universität Kiel) did a final check of the manuscript. Also, he accompanied the progress of my work with many discussions from the critical point of view of a theoretical physicist.

Thanks a lot to my brother Frank E. Westphal (Darmstadt) for drawing the overview of the Bahamas (figure 1). For the information on the origin of the name of the Bahamas I wish to thank Axel Pinck (Hamburg).

No study would be possible without discussions and exchange of experience in an agreeable working environment. Explicitly I want to mention discussions with Florian Böhm and Dierk Blomeier. Niels Andresen performed the calibration measurements of the X-ray diffractometer. Rebecca Rendle, as a native speaker, checked the English language. Ingo Gläser was a very capable research assistant. Also to the other colleagues: thanks for the good working atmosphere at GEOMAR!

For technical assistance at the SEM, I want to thank Werner Reimann (Universität Kiel), who does the best coatings of SEM stubs. Ute Schuldt (Universität Kiel) assisted with excellent photographical work. To GEOMAR I am grateful for providing the facilities. Especially I want to thank the librarian Angelika Finke, who was a patient and competent help.

Financially, this study was supported by the German Science Foundation (DFG Re 1051/3) and the Studienstiftung des deutschen Volkes. Also, the non-financial support of the Studienstiftung is gratefully acknowledged. The core CLINO was collected with funding from the United States National Science Foundation under Grants OCE 891-7295 and OCE 910-4294, from the Industrial Associates of the Comparative Sedimentology Laboratory of the Rosenstiel School of Marine and Atmospheric Sciences and from the Swiss National Science Foundation.

Last but not least, I wish to express my sincere thanks to my familiy who always accompanied my work with warm interesst and who supported me in numerous ways.

It is a pleasure to acknowledge the cooperation with Springer Verlag, especially with W. Engel during the preparation of this book.

Hildegard Westphal, Miami/Florida May 1998

Contents

Appendices   171

1 Introduction

1.1

Geology and Research History of the Bahamas

1.1.1
Recent Morphology

The Bahamian archipelago extends from southern Florida to the Puerto Rico Trench. It is located in 22°-28° N at the southeastern continental margin of the North American plate that formed during the Jurassic when Laurasia broke up and the North-Atlantic started to rift (Fig. 1). The modern topography of the Bahamas is characterized by two distinct realms; the shallow-water banks and the deep-water areas, that are separated by steep slopes.

The remarkable morphology of the Bahamas was already noticed by the Spanish explorers. Herrera (1601; in the English translation by Stevens, 1726) wrote about the reconnaissance of Ponce de Leon in 1513: "... *they went out from the islets, in lookout for Bimini, navigating among some islands he took to be overflowed and found it to be Bahama*". Craton (1992) remarks that "*it is an interesting fact that bajamar means shallow (strictly "low tide") in Spanish. Herrera was probably quoting the name given to the islands by the Spanish between 1513 and 1601*". The name of the Bahamian archipelago thus refers to a morphological description.

The shallow-water areas are composed of several steep-sided, flat-topped carbonate platforms. The shallow submerged platforms (<200 m water depth) sum up to 125,000 km^2 of the total area of the Bahamas (the latter amounts to approximately 300,000 km^2; Meyerhoff and Hatten, 1974). The Great Bahama Bank, located in the Southwest, represents the largest single platform of the Bahamas. Its shallow-water realm extends continuously over more than 400 km from north to south. The relief of the Bahama banks is low and most submerged areas are covered with less than 10 m of water (Newell, 1955; Newell and Imbrie, 1955). Small islands cap the banks mainly on their windward, eastern margins. They are composed of shallow-water reef material and lithified Pleistocene sand (Doran, 1955; Milliman, 1967). Their area sums up to about 11,400 km^2

(Meyerhoff and Hatten, 1974), an area similar to that of the state of Schleswig-Holstein/Germany that covers about 15,000 km^2 (Bibliogr. Institut Mannheim, 1974). Channels and re-entrants make up the deep-water realm. In Exuma Sound and Tongue of the Ocean, that cut into Great Bahama Bank, water depths exceed 1000 m. These deep-water areas are separated from the shallow-water platforms by steeply dipping slopes that generally show higher angles on the eastern (windward) side, and lower angles on the western (leeward) side of the platforms. Below the platform edge at 40-60 m below sea-level, the slopes dip almost vertically down to depths of around 135-145 m below sea-level ("the wall"; Grammer and Ginsburg, 1992; Grammer et al., 1993-a) At the base of this vertical wall, the slopes become less steep. Where facing the open Atlantic Ocean (Bahama Escarpment), the slopes reach depths greater than 4000 m below sea-level. With angles exceeding 40°, they belong to the steepest sustained modern continental slopes world-wide (Emiliani, 1965).

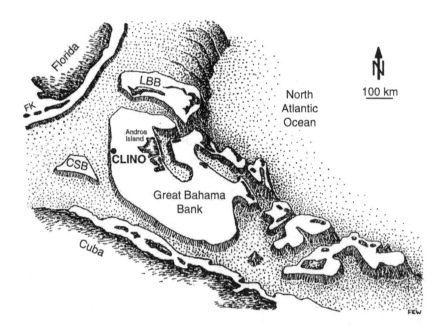

Fig. 1. Morphology of the Bahamas and the adjacent ocean basins. The Bahamian archipelago is located at the southeastern continental margin of the North American plate, at the edge of the North Atlantic ocean basin. Note the large areal extent and the low relief of the Great Bahama Bank. At the present day, most of the carbonate platform is covered by shallow water. *LBB*: Little Bahama Bank, *CSB*: Cay Sal Bank, *FK*: Florida Keys. Drawing by F. E. Westphal after various sources (e.g. Pilkey and Rucker, 1966).

This juxtaposition of extreme morphologies and the huge size of the Bahamas has been fascinating to geologists for a long time and the question of the origin of the Bahamas has served as subject of long-lasting debates.

1.1.2
Origin of the Bahamas

The Bahamas are among the most extensively studied regions in the world, and a number of phenomena typical of calcareous environments have been described first from the Bahamas. Geological research on the Bahamas has been documented first by a study of Nelson (1853) that describes the geography and topography of the Bahamas. He noticed the *"remarkable lowness of profile"* (Fig. 2) and the dynamics of construction and destruction of the islands. Also he outlined the biota and lithologies, and described the formation of the carbonate rocks he found. He was the first to notice the eolian origin of many Bahamian islands. The examination of modern carbonate environments experienced a rapid progress with the expedition of L. and A. Agassiz to the Bahamas in 1893 (Agassiz, 1894). Their explorations focused mainly on the fringing reefs of the Great Bahama Bank. Research on abiogeneous carbonate components followed, e.g. by Vaughan (1910). Vaughan emphasized that carbonate constituents can originate from skeletal secretion or from chemical precipitation and introduced the terms "organic" and "inorganic" limestones.

In spite of the long-lasting research history of the Bahamas, until recently the origin of the Bahamas has been controversially debated (Mullins and Lynts, 1977). Among the questions discussed are firstly the continental or oceanic nature of the basement of the Bahamas, and secondly the origin of the morphology of the Bahamas.

1.1.2.1
Basement of the Bahamas

Since the geological studies of Nelson (1853), at least 36 theories on the origin of the Bahamas have been discussed (see summaries in Meyerhoff and Hatten, 1974 and Mullins; 1975). Early authors thought that the Bahamas were underlain by continental crust: Nelson (1853) assumed the Bahamas to be a huge delta formed by the Gulf Stream. Hilgard (1871, 1881), Gabb (1873), and Suess (1888, 1908) thought the Bahamas to be isolated fragments of the Gulf of Mexico-Florida tectonic plate composed of Tertiary carbonates. Despite such differences in the details, the general concept of an underlying continental crust has long been accepted.

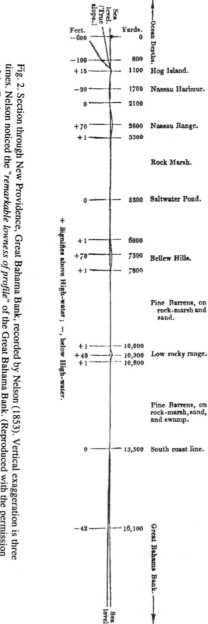

Fig. 2. Section through New Providence, Great Bahama Bank, recorded by Nelson (1853). Vertical exaggeration is three times. Nelson noticed the "*remarkable lowness of profile*" of the Great Bahama Bank. (Reproduced with the permission of the Geological Society London.)

In the late 1960s and early 1970s, the concept of plate tectonics questioned this hypothesis of a continents basement of the Bahamas. In a reconstruction of the pre-rift continents by Bullard et al. (1965), the area of the Bahamas entirely overlaps the African continent. Based on this observation the hypothesis formed that the Bahamas were underlain by oceanic crust (Dietz et al., 1970; Le Pichon and Fox, 1971; Glockhoff, 1973; Sheridan, 1974). Meyerhoff and Hatten (1974), nevertheless, supported their original assumption of an underlying continental crust by geophysical data and by the geological similarity to the adjacent areas of Yucatan and Florida. Their concept finally was confirmed by Mullins and Lynts (1977) who resolved the overlap problem by a pre-rift reconstruction with rotation of the region involved, thus obtaining a perfect fit.

One of the objectives of the first deep subsurface research on the Bahamas was to gain information on the basement. Four borings were drilled in the 1970s, three of them being located on Great Bahama Bank (Andros-1, total depth 4,448 m; Long Island-1, 5,355 m; Great Isaac-1, 5,443 m), and one on Cay Sal Bank (Cay Sal IV-1, 5,766 m; Meyerhoff and Hatten, 1974; Schlager et al., 1988). The oldest rocks recovered were Jurassic carbonates and black shales in Great Isaac-1. The Cretaceous sediments present in all four wells are composed of shallow-water carbonates. The total thickness of Cretaceous to Tertiary sediments of roughly 5000 m indicated enormous subsidence rates. These wells, however, did not reach the basement, thus leaving the question of the underlying lithology unanswered.

1.1.2.2
Origin of the Morphology

Connected to the question of the nature of the underlying crust, the second long-lasting controversy dealt with the origin of the striking present-day morphology of the Bahamas. Three general concepts are proposed in the literature: (a) the concept of an inherited tectonic structure, (b) the concept of an inherited erosional morphology, and (c) the concept of a constructional Cretaceous carbonate system.

(a) It was long believed that the recent morphology reflects an underlying, buried relief. This precursor morphology was interpreted to be inherited from basement structures like folds or faults (e.g. Talwani et al., 1960; Ball et al., 1969; Lynts, 1970; Sheridan, 1971, 1974, 1976; Uchupi et al., 1971; Glockhoff, 1973; Mullins and Lynts, 1977). These fault structures were related to the Jurassic rift event (Mullins and Lynts, 1977). Even a volcanic precursor topography was advocated (Schuchert, 1935).

(b) Hess (1933, 1960) assumed a subaerial erosional relief like a drainage pattern as precursor morphology. Ericsson et al. (1952), Gibson and Schlee (1967), and Andrews et al. (1970) suggested that the deep Bahama channels are essentially valleys eroded in the submarine environment by slumping and fluxo-turbidity currents.

(c) The concept of a constructional origin of the morphology was based on a comparison of the steep marginal profiles with the vertical growth of Pacific atolls (Newell, 1955). Dietz et al. (1970), Paulus (1972), and Dietz and Holden (1973) thought that the present-day morphology is inherited from reef development in the Lower Cretaceous. The so-called "megabank" concept states that the modern patterns result from a disintegration of a much larger carbonate platform in the Cretaceous (Paulus, 1972; Schlager and Ginsburg, 1981; Sheridan et al., 1981). This megabank is thought to have drowned during the Mid-Cretaceous. According to this concept, the recent platforms represent isolated re-established platforms (Schlager and Ginsburg, 1981).

In contrast to the three aforementioned concepts that assume an upward growth after the establishment of the platform without considerable lateral migration of the margins, Ball (1967, 1972) noted that the basement structure probably did not contribute much to vertical relief. He stated that any precursor structure most likely has significantly been modified by vertical growth. The existence of a Cretaceous megabank has been corroborated by ODP Leg 101 (Austin et al., 1988), and thereby an end has been put to the debate of the underlying lithology. Additionally, the results of ODP Leg 101 and a re-evaluation of core Great Isaac-1 revealed that in contrast to the more static concepts, some margins of the Northwest Bahamas have migrated laterally over 10 km since the Miocene (Schlager et al., 1985; Austin et al., 1986; Austin et al., 1988; Schlager et al., 1988).

Later, a high-quality seismic reflection profile across Great Bahama Bank, that reached depths of 1.7 seconds two-way travel time (depth of penetration >2500 m), considerably improved the understanding of the development of the Bahamian morphology (Fig. 3; Eberli and Ginsburg, 1987, 1989). They showed that the modern topography is the result of much more dynamic processes even than inferred from ODP Leg 101, including significant bank migration. The modern topography thus does neither reflect rift-graben structures nor the morphology of a Cretaceous megabank. The modern Northwest Great Bahama Bank evolved from a process of repeated tectonic segmentation, probably related to reactivation of tectonic movements associated to the collision of the North American plate with Cuba. Progradation led to subsequent coalescence. On the seismic profile, Eberli and Ginsburg, (1987) identified two linear north-south trending subsurface depressions that in the Mid-Cretaceous and Tertiary separated three smaller banks. These depressions afterwards coalesced by lateral accretion. Also, lateral growth on the leeward side of Great Bahama Bank led to progradation of more than 25 km westward into the present-day Straits of Florida. Since the Cretaceous, the repeated tectonic segmentation and subsequent coalescence, induced by the high productivity of the shallow-water carbonate factory, led to the progressive modification of the Great Bahama Bank. Thus, progradation is one of the most striking features governing the development of this carbonate platform.

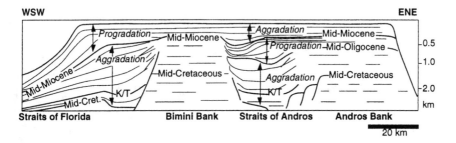

Fig. 3. Interpretation of seismic section ("Western Line") through Great Bahama Bank displaying the complicated internal architecture of the bank. Two nuclear banks, Andros and Bimini Banks, coalesced by the infilling of an intraplatform seaway, the Straits of Andros. Progradation of the western margin of the platform during the Neogene expanded the bank more than 25 km into the Straits of Florida (Modified from Eberli et al., AAPG©1994, reprinted with the permission of the American Association of Petroleum Geologists).

1.1.3
Ramp versus Steep-Sided: Morphology Evolution and Sea-Level

Interpretation of the Western Line (Figs. 3 and 4) also led to the realization that the morphological development of Great Bahama Bank at least from the Miocene on, is closely linked to the glacio-eustatic sea-level history (Eberli and Ginsburg, 1987, 1989, Pomar, 1993). For the late Tertiary and the Quaternary, seismic sequences as well as morphology modifications can be related to sea-level events.

Eberli et al. (1997) defined 9 seismic sequences (*a* to *i*) in the Miocene to recent succession (Fig. 4; seismic sequence interpretation summarized here follows Eberli and Ginsburg, 1987, 1989; and Eberli et al., 1997). Progradation in the upper Miocene (sequences *i, h, g,* Fig. 4) indicates that little accommodation space was created during this time. The Western Line shows a Miocene platform with a distinct break in the slope to the ENE of CLINO at the top of sequence *h*. The Miocene-Pliocene boundary (seismic sequence boundary *g/f*) coincides with a global sea-level fall associated with the expansion of the Antarctic ice sheet (Kenter et al., in press). This sea-level fall was correlated with a 50 m drop on the global eustatic sea-level curve of Haq et al. (1988).

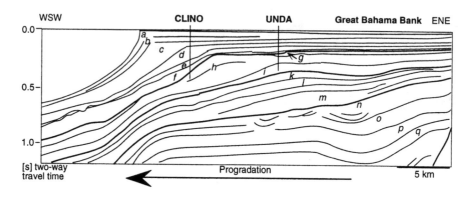

Fig. 4. Detail of the Western Line showing the leeward clinoforms that prograde westward from the Bimini Bank. The cores CLINO and UNDA of the Bahamas Drilling Project that is described in Chapter II.1, penetrate the Miocene (*i* and *h*) to recent (*a*) prograding succession of slope sediments. Intervals selected from CLINO for the present study are in seismic sequences *f* and *d*. From Eberli et al. (in press).

The subsequent high-amplitude sea-level rise resulted in temporary drowning of the carbonate platform (Eberli et al., 1997; Kenter et al., in press). Basinal lowstand geometries are observed on top of the sequence boundary. The major backstepping of the platform margin resulted in a change of the morphology of Great Bahama Bank. In the Lower Pliocene, the flat-topped platform that prevailed during the Miocene, developed into a ramp with an atoll-like morphology. The early Pliocene ramp was characterized by slope angles between the platform margin and CLINO of around 3.5° at sequence *f*. The slopes of the carbonate platform were thereby somehow steeper than those attributed to a ramp in a strict sense, where low-gradient slopes do not exceed 1° (Ahr, 1973). The Lower Pliocene ramp rather corresponds to a distally steepened ramp (Read, 1982, 1985). Burchette and Wright (1992) pointed out that from a sedimentological point of view, a distally steepened ramp resembles rather a homoclinal ramp than a rimmed, flat-topped platform. The term "ramp" therefore here is used in its wider sense to describe the Lower Pliocene morphology of the leeward slope of Great Bahama Bank.

When the earliest Pliocene overall sea-level rise slowed down, the platform was able to aggrade to sea-level and subsequently to prograde over a large distance, creating again a flat-topped morphology (sequences *f*, *e*, *d*; Figs. 4 and 5). The upper Pliocene shows prograding geometries and a steepening of the flanks of this flat-topped platform. Between the shelf break and CLINO, the flanks show a dip of about 6° at the base of sequence *d*. At the top of the sequence, the platform margin has prograded over the location of CLINO, and the angle between UNDA and CLINO is almost horizontal. In Pliocene-Pleistocene platform top sediments, Beach and Ginsburg (1980) and Beach (1982) observed a change in composition from a dominance of skeletal to non-skeletal grains. This change is interpreted to reflect the morphologic evolution from an open, atoll-like bank with gentle slopes, into a flat-topped bank, that resembled the modern morphology (Fig. 5; see also Schlager and Ginsburg, 1981; McNeill et al., 1988; Reijmer et al., 1992). The upper, dominantly non-skeletal formation represents a major change in sediment deposition due to a changed circulation and shallowing of the bank (Beach and Ginsburg, 1980).

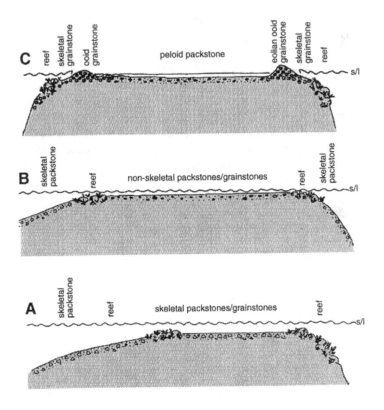

Fig. 5. Schematic morphologic evolution of Great Bahama Bank from the Pliocene to the Upper Pleistocene. During the Lower Pliocene (**A**), Great Bahama Bank was characterized by a distally-steepened ramp morphology. Steepening of the flanks due to progradation resulted in the flat-topped, steep-sided morphology of the Upper Pliocene platform (**B**). Further steepening in the Pleistocene (**C**) finally resulted in present day morphology (not shown) that is typified by a virtually vertical wall down to depths of 140 m below sea-level (modified after Reijmer et al., 1992, reprinted with the permission from Elsevier Science).

The non-skeletal sediments have been termed "Lucayan Limestone" and are dated as late Pliocene-Pleistocene (time-equivalent to the Floridan Key Largo Limestone and Miami Limestone; Beach and Ginsburg, 1980). The Lucayan Limestone is bounded above by a Late Pleistocene unconformity that forms the present-day surface of many islands along the margin of the banks, and underlies unlithified Holocene sediments in the interior of Great Bahama Bank. This change in depositional style from skeletal to non-skeletal observed in the platform interior is also reflected in adjacent toe-of-slope turbidites from ODP Leg 101 (Reijmer et al., 1992).

The high-frequency oscillations, caused by the onset of the glaciation of the Northern Hemisphere during the Plio-Pleistocene resulted in sea-level rises overstepping the platform top only slightly. Export of shallow-water material resulted in pronounced progradation of the uppermost Pliocene to Quaternary (sequences *c, b, a;* Fig. 4), and the slopes of the platform continued to steepen (Eberli et al., 1997).

Although progradation has been recognized as a most important feature in the development of carbonate platforms, the mechanisms of progradation are not yet fully understood. To gain a better understanding of progradation, a thorough comprehension of periplatform-sedimentation on the slopes of carbonate platforms is required. For this aim, the Bahamas Drilling Project, that is described in Chapter 2.1, was initiated (Ginsburg, in press-a). The present study, that deals with samples from the Bahamas Drilling Project, aims to contribute to the discussion on slope sediments by examining smaller-scale variations in periplatform sediments.

1.2

Objectives

The objective of this study is to examine variations in the composition of periplatform sediments with respect to the morphologic evolution of a carbonate platform. The sedimentary signature of sea-level variations in slope sediments is shown to depend on the flat-topped versus ramp-shaped morphology. Subsequently, the diagenetic signatures observed in these periplatform sediments are related to the input patterns.

Due to the dependence on light of most carbonate-secreting organisms, they build up close to sea-level (Schlager, 1981), frequently forming flat-topped carbonate platforms. On these flat-topped platforms, even low-amplitude sea-level fluctuations commonly result in repeated exposure and flooding of the platform top. This strongly influences the carbonate production on the platform (Schlager, 1981; Droxler and Schlager, 1985). Subaerial exposure of the platform top drastically cuts down the production of sediment in the shallow-water carbonate factory. Sediments deposited on the top of a carbonate platform are usually characterized by numerous exposure surfaces. Optimum conditions for high production prevail when the rate of sea level rise corresponds to the potential rate of carbonate production (Kendall and Schlager, 1981). The influence of sea-level fluctuations on the morphologic and sedimentologic development of flat-topped carbonate platforms has been described from numerous carbonate platforms of different ages (e.g. Jurassic, Morocco: Crevello, 1991; Triassic, Dolomites: Bosellini, 1984; Goldhammer et al., 1990; Permian, New Mexico: Silver and Todd, 1969; Sarg, 1988; Devonian, Canning Basin: Kennard et al., 1992).

Flooded carbonate platforms like the present-day Little Bahama Bank produce up to three times the amount of carbonate mud as can be stored within the present-day shallow-water area (Neumann and Land, 1975). The excess sediment is exported to the periplatform realm (Neumann and Land, 1975). These periplatform sediments are predominantly composed of mud and larger grains produced on and exported from the platform top, rim and foreslope, that are mixed with planktic input (Schlager and James, 1978). As sediment export during sea-level highstands, when the carbonate factory is submerged, by far exceeds the export to the slopes during lowstands, periplatform records are usually characterized by higher sedimentation rates during highstands (highstand shedding *sensu* Droxler and Schlager, 1985). Depending on a variety of factors such as accommodation space, growth rates, and currents, the varying export of sediment towards the flanks may record the flooding and exposure events experienced by the shallow-water "carbonate factory". Contrary to platform-top successions, under optimum conditions (no to moderate winnowing) slope sediments may provide a rather continuous sedimentary record of the changes in the shallow-water productivity. Slope sediments thus can be employed to study environmental changes affecting the shallow-water carbonate factory by deciphering their compositional signatures (e.g. Droxler and Schlager, 1985; Schlager and Camber, 1986; Goldhammer and Harris, 1989; Haak and Schlager, 1989; Brachert and Dullo, 1991, 1994; Grammer et al., 1991; Grammer and Ginsburg, 1992; Harris, 1994). Grammer and Ginsburg (1992) have shown that on the Bahamas, Holocene highstand deposits consist of thick wedges of fine-grained carbonate sands and muds that are derived from the platform top (compare Wilber et al., 1990), whereas thin talus deposits from the preceding lowstand are composed of coarse material eroded from the steep wall (see also Schlager and Camber, 1986).

The concept of highstand shedding, however, mainly applies to flat-topped carbonate platforms with high-angle slopes (Schlager et al., 1994). Ramps react to sea-level fluctuations in a different way than flat-topped platforms (Burchette and Wright, 1992). On many carbonate ramps which are characterized by low-angle slopes and the absence of a pronounced edge, facies belts are forced to shift downslope during minor sea-level falls ("forced regression" of Posamentier et al., 1992) rather than being "switched off" as the shallow-water factory of a flat-topped platform (Burchette and Wright, 1992; Bachmann and Willems, 1996). This results in smaller changes in composition between lowstand and highstand sediments deposited on the slope. Burchette and Wright (1992) noted that the periplatform record of carbonate ramps, in comparison to flat-topped platforms, is largely uninvestigated. Since this statement, research on periplatform facies of fossil ramps has increased (e.g. Elrick and Read, 1991; Bachmann et al., 1996; Bachmann and Willems, 1996), but still many questions remain subject to discussion. Among those are many aspects of the sequence stratigraphic expression on a small scale (microfacies successions), and likewise the facets of diagenetic alterations of slope sediments of a carbonate ramp. To contribute to the discussion, the present study focuses on the signatures of sea-level fluctuations of the order of 100 ka (4th order of Vail et al., 1991) in periplatform sediments of a flat-topped platform in comparison with a ramp-type configuration. Because of the Pliocene morphologic evolution, the periplatform sediments of Great Bahama Bank offer the opportunity to investigate slope sediments of both morphology types at the same locality.

This study addresses two major questions: Firstly, the compositional input patterns observed in periplatform sediments and their relationship to sea-level fluctuations and the evolution of platform morphology, and secondly the early diagenesis of these periplatform sediments with respect to their varying composition. The study is structured as follows: Chapter I (Introduction) gave a short description of the recent morphology of the Bahamas, and a reminiscence of the research history and the controversially discussed questions regarding the origin of the Bahamas. In Chapter 2 (Study Area and Material), the Bahamas Drilling Project, that led to the recovery of cores CLINO and UNDA, is briefly presented, and the material from core CLINO that was investigated in the present study is described. In Chapter 3 (Methods) the procedures are described that were used to address the objectives of this study. In Chapter 4 (Primary Signals - Sediment Input), the focus will be on the compositional signatures observed. First the results of the compositional analysis are presented, then they will be discussed in the context of sea-level and morphologic evolution. In Chapter 5 (Secondary Signals - Diagenesis), the diagenetic patterns will be described. After presenting the results of mineralogical, geochemical, and petrographic examinations, the features are interpreted with respect to compositional signatures, and a lithification model is proposed.

2 Study Area and Material

2.1

The Bahamas Drilling Project

In order to obtain a better understanding of the mechanisms of periplatform sedimentation and progradation, the Bahamas Drilling Project was designed (Fig. 6; Ginsburg, in press-a). Two boreholes, CLINO and UNDA, were projected to penetrate the prograding Tertiary paleo-slope of Northwestern Great Bahama Bank, facing the open seaway of the Florida Straits. The names of the two boreholes, that are located in the recent "clinoform" (slope) and "undaform" (platform top) environments of Rich (1951), allude to the nomenclature of this author. To obtain an optimum stratigraphic control, the two drill sites were located on the aforementioned seismic line ("Western Line"; Fig. 4). The Bahamas Drilling Project mainly aimed to address sequence stratigraphic questions such as facies information for the seismic sequences, nature of progradational facies, role of sea-level in controlling progradation, petrophysical causes of seismic reflectors, and age control for the seismic sequences (Ginsburg, in press-b; Eberli et al., 1997). Drilling took place between March and May 1990. Conventional diamond drilling technique was carried out from a self-propelled jackup barge.

Since then, a number of petrographic, sedimentologic, geochemical, and petrophysical studies were carried out on the sediments recovered during the Bahamas Drilling Project, most of which are being published together in an SEPM Contributions in Sedimentology (Ginsburg, in press-a: Melim et al., in press-a, in press-b; Ginsburg, in press-b; Ginsburg et el., in press; Swart et al., in press-a, in press-b; Eberli et al., in press; Kenter et al., in press; see also: Melim et al., 1995, Melim, 1996).

The western, leeward slope of Great Bahama Bank is characterized by low-angle prograding clinoforms of Tertiary to Quaternary age that form a succession that shallows upward from lower to upper slope facies (Eberli and Ginsburg, 1987; 1989). The Western Line revealed that the seismic sequences are laterally stacked (Fig. 4). This pattern is thought to be sea-level induced as downstepping of the margins could be identified. Thus, sea-level could potentially be recorded in the seismic section (Eberli and Ginsburg, 1987, 1989; Eberli et al., 1997). The Bahamas Drilling Project cores penetrated the proximal portions of the prograding seismic sequences seen on the Western Line (Eberli et al., 1997). UNDA is located on the recent platform top in 6.70 m of water depth and aimed to penetrate an upper-slope to shallow-water succession with platform margin deposits. CLINO is positioned 8.5 km further to the southwest in a water depth of 7.60 m at the recent margin of the platform. It was designed to penetrate a series of inclined seismic reflectors that are thought to image slope deposits underneath a thin upper Pleistocene and Holocene platform succession (Eberli et al., 1997). CLINO recovered a succession of Neogene slope sediments and reached upper Miocene sediments (seismic sequence *h*) at a total depth of 677.27 m below mud pit (mbmp; mud pit is 7.30 m above sea-level. In the following, depth information regarding CLINO will always be given in mbmp). Recovery was 77 % for hole CLINO. The slope sediments consist of a mixture of platform-derived and open-marine skeletal and non-skeletal constituents. Grain sizes vary from mud to sand in these very pure carbonates (Kenter et al., in press).

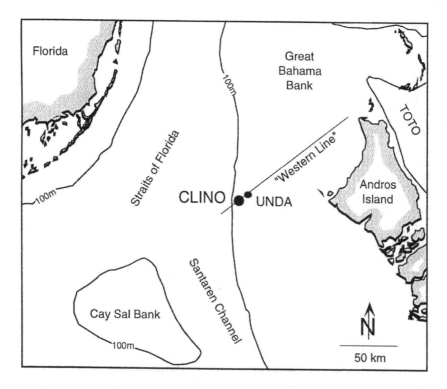

Fig. 6. Location of the Bahamas Drilling Project cores CLINO and UNDA on Great Bahama Bank. Core CLINO, that has been sampled for the present study, is located on the leeward present day platform margin. *TOTO* = Tongue of the Ocean (from Eberli et al., 1997).

Lithologic examinations of CLINO led to the recognition of four large-scale depositional sequences that are separated by distinct unconformities (Fig. 7; Kenter et al., in press). The lower two of these unconformities could be correlated to the seismic sequence boundaries separating seismic sequences *h* and *f* (seismic sequence *g* was not recovered), and sequences *e* and *d* (Eberli et al., 1997; Kenter et al., in press). Dating of the cores was done by integrating biostratigraphic, magnetostratigraphic, and strontium isotope data (Lidz and McNeill, 1995-a, 1995-b; Swart et al., in press-a; for a summary see Eberli et al., 1997). As seen in UNDA, on the platform top, seismic sequence boundaries can correspond to subaerial exposure surfaces or changes in facies, whereas on the slope (CLINO), sequence boundaries are expressed by distinct discontinuity surfaces. These discontinuity surfaces are characterized by hardgrounds and mostly are overlain by blackened lithoclasts (Eberli et al., 1997). An unconformity at 536.33 mbmp separates (IV) Miocene and (III) Lower Pliocene deposits and is thought to coincide with the globally observed drastic expansion of the Antarctic ice sheet and the associated sea-level fall. The lowermost Pliocene sediments (536.33-507.70 mbmp) are characterized by a predominance of planktic foraminifers. They are interpreted to have settled from the water column with minor contribution from the margin (Kenter et al., in press). Above this interval, the lower Pliocene deposits are dominated by fine-grained skeletal and non-skeletal sediments. The (III) Lower and (II) Upper Pliocene deposits are separated by an unconformity at 367.03 mbmp. This boundary correlates with the onset of the first post-Miocene cooling (Kenter et al., in press). The (II) Upper Pliocene to (I) lowermost Pleistocene sediments are characterized by fine-grained components. Between the lowermost Pleistocene and the remaining Quaternary deposits, an unconformity is observed at 197.44 mbmp. Above this unconformity, shallow-water and reefal deposits dominate. Overall, the Pliocene to recent lithology of CLINO reflects a shallowing upwards trend that corresponds to the progradation of the slope across the locality of the drilling site. Internally the three large-scale depositional sequences are composed of subordinate cycles in the order of 10's of meters, that in turn show an internal smaller-scale cyclicity (Kenter et al., in press). The nature of the smaller-scale variations that are represented by changes in grain sizes and degree of cementation is investigated in the present study.

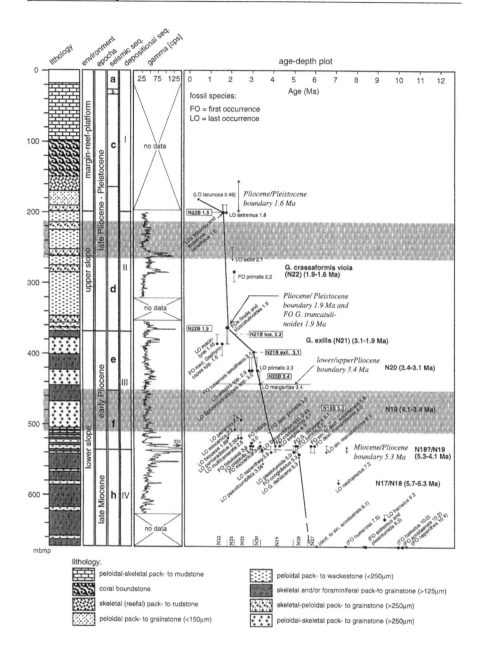

Fig. 7. Core CLINO. For caption see opposite page.

Fig. 7. Core CLINO. The two intervals examined in the present study cover part of the Upper Pliocene biozone N22 and part of the Lower Pliocene biozone N19 (shaded intervals). Lithological logs, interpretation of depositional environments, and gamma-ray curve from Kenter et al. (in press), seismic sequences and (simplified) depositional sequences from Eberli et al. (1997), and age-depth relationship and dating from McNeill et al. (in press).

2.2
Sampling and Material examined

The present study concentrates on Pliocene sediments from the CLINO core which were deposited in middle to upper slope environments. Sampling took place in March 1995 and September 1996 at the Rosenstiel School of Marine and Atmospheric Sciences (University of Miami) where the Bahamas Drilling Project cores CLINO and UNDA are stored. High-resolution sampling (15-30 cm distance between samples) of Pliocene successions from the core CLINO has been undertaken in order to obtain an insight in the aforementioned subordinate levels of cyclicity in periplatform carbonates. To assess the effects of changes in platform morphology (ramp versus steep-sided) on the composition of periplatform carbonates, successions from two different large-scale sequences have been included in this study. An interval of Upper Pliocene age (seismic sequence *d*) was selected to represent a steep-sided carbonate platform, whereas an interval of Lower Pliocene (seismic sequences *f*) was chosen to represent a distally-steepened ramp (Figs. 4 and 7). Both intervals appeared suitable for a study of the subordinate cyclicities, as they exhibit macroscopically visible variations in grain-size, color, and porosity. The two intervals also seemed appropriate for this study, because the gamma-ray log exhibits one conspicuous cycle in the selected Upper Pliocene interval, and three well-developed saw-tooth shaped cycles in the selected Lower Pliocene interval. The intervals sampled generally are characterized by white to creamy colors, where laminations and bioturbation are difficult to distinguish due to low color contrasts. Most parts of the core that have been sampled are macroscopically rather featureless fine-grained carbonates, some intervals are conspicuous by a coarser, biodetrital composition.

The upper Pliocene succession (in foraminifera zone N22, 217.04 to 264.11 mbmp, 182 samples; Fig. 7; Appendix 1) was sampled at distances of 30 cm between the samples. Parts that appeared interesting by a more variable composition have been sampled at smaller intervals. This selected upper Pliocene interval represents slope sediments mainly derived from a rimmed, flat-topped platform. It consists of one macroscopically visible cycle with two thinner layers of coarse-grained material bounding a thicker interval of fine-grained material.

The lower Pliocene succession (in foraminifera zone N19, 451.71 to 509.47 mbmp, 287 samples; Fig. 7; Appendix 1) was sampled at distances of 15 cm. Locally, denser or wider sampling was performed. The upper 15 m of this interval were only sampled sporadically. The selected Lower Pliocene interval was deposited during times when a ramp morphology prevailed. It is composed of three subtle, macroscopically distinguishable cycles with thicknesses in the order of 10's of m. They are defined by wackestone to packstone alternations in which grain size does not differ significantly. At the base of the selected interval, some samples of the coarser-grained deposits were taken that characterize the lowermost Pliocene.

Below 145 mbmp (thereby including the two intervals examined), the core CLINO offers the opportunity to investigate the sedimentary record of slope carbonates beneath the reach of meteoric diagenetic fluids (Melim et al., 1995; Melim, 1996). Thus, it is possible to investigate lithification processes that are assumed to be typical for periplatform carbonates, because these deeper-water slope carbonates usually experience non-meteoric early diagenesis.

3 Methods

3.1
Compositional Analysis

3.1.1
Point Counting

Description and determination of the components present in the samples is based mainly on thin section analyses, but scanning electron microscopic (SEM) results are also included (see Chapter 3.4).

A quantitative component analysis has been undertaken by point-counting thin sections using the method of Chayes (1956). A total of 369 samples has been analyzed. In each thin section, 200 points have been counted to obtain a reasonably low statistical error (Van der Plas and Tobi, 1965). The samples were counted volumetrically (Flügel, 1978), i.e. individual grains are counted several times if they occupy several points of the grid. Internal pore space and internal cements are also counted as being part of the surrounding bioclast.

The quantitative component analysis aims at differentiating the various sediment sources of the periplatform sediments (platform interior, rim, slope, open marine). This method has been shown to be efficient for toe-of-slope and slope sediments where the specific signatures of different production zones were recognized (Haak and Schlager, 1989; Everts, 1991; Reijmer and Everaars, 1991; Reijmer et al., 1991; Everts and Reijmer, 1995). To resolve which facies zones were involved during the deposition of the different samples, characteristic groups of components have been distinguished that are thought to represent specific paleoenvironments (Fig. 8A; based on e.g. Flügel, 1978; Enos, 1983; Haak and Schlager, 1989; Ginsburg et al., 1991; Burchette and Wright, 1992; Everts and Reijmer, 1995). The same groups have been considered for both, flat-topped platform and ramp morphologies. The distribution and widths of the production zones, however, differ slightly between the different morphologies (Fig. 8B). Also, on ramps, the transitions between the different production belts are broader and less well defined.

The following groups were distinguished (for a detailed description of the components and for references to the literature see Chapter 4.2.1.):

(A) Skeletal grains:
(1) *Halimeda* plates that originate predominantly from the rim and upper slope. (2) Red algae that are represented by nodules and branches, also derived from rim to upper slope environments. (3) Mollusks that are dominantly shed from the platform top. (4) Bryozoans that are rim to slope inhabitants. (5) Benthic foraminifers, mainly shallow-water species such as Amphisteginidae, Miliolidae, and specific Rotaliidae, and deeper-water forms such as some Textulariidae. (6) Planktic foraminifers, mainly Globigerinidae, rarely Globorotaliidae, that represent the open ocean source environment. Pteropods, that also originate from the open ocean are extremely rare. (7) Ostracods that are non-diagnostic with respect to their source environment. (8) Echinoderms that are non-diagnostic, because in the samples examined, forms occurring on the platform top and rim usually can not be differentiated from forms occurring in deeper water. (9) Non-determinable grains: The objective was to quantify the composition of the predominantly allochthonous sediment at the time of deposition on the slope. Therefore diagenetically altered grains (molds, neomorphoses, small-sized biodetritus) have been treated as their primary precursors when recognizable. Otherwise they have been incorporated into the category of non-determinable grains.

(B) Non-skeletal components:
(10) Cortoids that originate mainly from sandy shoals. (11) Intraclasts that have been redeposited when still unlithified as is indicated by their smooth boundaries. (12) Fine-grained material, constituting the matrix, that is thought to be derived originally from the platform interior. (13) Peloids that originate mainly from the platform interior. The predominantly fecal origin of the pellets is indicated by tiny burrows, that are characteristic for soft-material burrowers (coprophags). In thin sections of strongly recrystallized samples, however, homogenized grains sometimes could not be distinguished from peloids of presumably fecal origin. (14) Primary interparticle pore space (generally preserved as sparry cement).

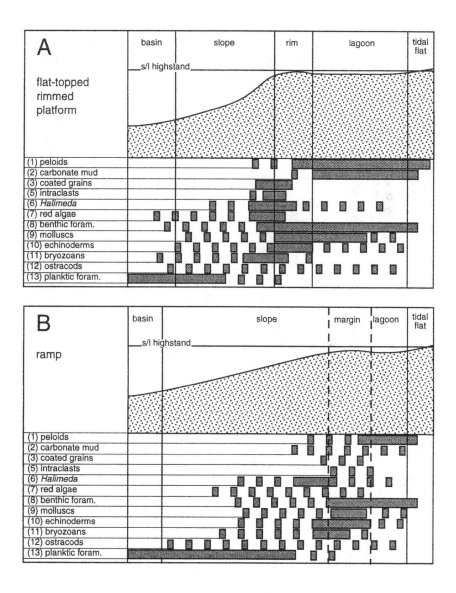

Fig. 8. Distribution of the facies belts sourcing the constituents distinguished in point-counting. (**A**) Source areas on a steep-sided, flat-topped platform. (**B**) Source areas on a distally steepened ramp. Based on Flügel (1978), Enos (1983), Haak and Schlager (1989), Ginsburg et al. (1991), Burchette and Wright (1992), and Everts and Reijmer (1995).

Numerical analyses have been applied to quantify the compositional variations in the intervals examined. Therefore, the results of the compositional analysis (point counting) were further analyzed by different statistical methods. Frequency distribution and basic summary statistics were calculated to estimate the significance of point-count groups in characterizing specific deposits and the variance of each point-count group within the two intervals. The statistical similarity between the individual groups was assessed using Pearson's correlation coefficient analysis. Principal components were calculated from the raw data to determine which groups steer the characterization of the point-count data set. Factors as linear combinations of related variables (dimensionality reduction) are based on eigenvalue-eigenvector analyses. Numerical methods have been applied to similar data sets by Reijmer and Everaars (1991), Reijmer et al. (1991, 1992), and Bachmann and Willems (1996). For a detailed description of these methods the reader is referred to Davis (1973) and Rock (1988).

For statistical analyses, all samples were included that have been point-counted. Nevertheless, in the selected Lower Pliocene interval, four duplicate samples (B-samples) were excluded from statistical analyses, thus leaving a data set of 238 samples. In the Upper Pliocene succession, in two intervals of special interest, significantly denser sampling was performed (5-10 cm distance, in contrast to the usual 30 cm). To obtain an approximately equidistant data set, i.e. for not overemphasizing the more densely sampled intervals, the additional 24 samples were excluded, thus leaving a data set of 155 samples (see Appendix 1).

3.1.2
Quantitative Analysis of Foraminiferal Associations

The abundance of benthic and planktic foraminifers within thin sections has been determined by counting the individuals present in an area of 1 to 9 cm^2 depending on the homogeneity of the thin section. Thus, statistically significant numbers were achieved that could otherwise not be obtained by point-count analysis due to the small size of most individuals (and thus their small areal percentage in the thin sections). Foraminiferal assemblages were determined in order to extract environmental information on the source area with respect to restriction and salinity, and on the facies belts productive during the time of deposition.

In contrast to loose samples, foraminifers in lithified carbonate rocks are more difficult to examine as less specimens are found in thin section than in sieved samples. Also, most specimens in thin section are not cut properly oriented, inhibiting a precise determination (Rose and Lidz, 1977). Additionally, in thin sections there is an inherent bias against small species that frequently are crushed, dissolved or recrystallized (if they have thin tests), or are overlooked or mistaken for juvenile forms of other species. Therefore, only clearly morphologically separable groups are distinguished (Fig. 9). The environmental interpretation of these groups is based on studies by Murray (1973), Rose and Lidz (1977), and Lidz and McNeill (1995-a):

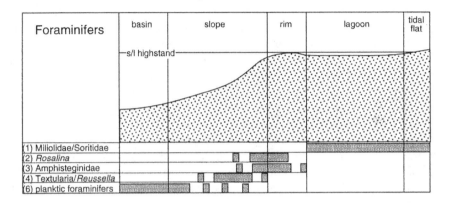

Fig. 9. Foraminifer groups typical for specific environments. The distribution on the present-day steep-sided, flat-topped platform is shown. Compiled from Murray (1973), Rose and Lidz (1977), and Lidz and McNeill (1995-a).

(A) Platform-top: (1) Miliolidae are interpreted as indicators of shallow water. Rotund forms are indicative for extremely shallow water. Soritidae similarly are indicative for shallow water.

(B) Margin: (2) *Rosalina* is found in outer shelf sediments and is closely related to the presence of reefs. (3) Amphisteginidae have a very limited distribution and only flourish in outer-margin and shallow-slope environments.

(C) Slope: (4) Textulariids and *Reussella* are typical for deeper settings, and are easily distinguishable even in thin section.

Benthic groups with insufficient individuals, non-diagnostic groups, and indeterminable specimens are included in the group (5) "other benthic foraminifers". Among others, they include mainly different Rotaliidae, but also Elphiidae, and Eponidae.

(D) Open ocean: (6) Pelagic forms like Globigerinidae and Globorotaliidae occur in slope to basinal sediments and are rare in outer-margin deposits.

3.1.3
Palynologic Examinations

Organic-walled microfossils were examined *in situ* in polished, slightly etched bulk rock samples with the SEM, applying the method of Munnecke and Servais (1996). This method is suitable to obtain information on the preservation (especially possible deformation) of the thin-walled organic microfossils (Westphal and Munnecke, 1997). For a detailed description of the SEM method, see Chapter 3.4.

Additionally, palynologic determinations have been carried out by M. Head, University of Toronto. Where possible, the palynomorphs were determined to species level. For these palynological determinations, the samples were treated with cold HCl and HF. Oxidants and hot acids were not used since these can damage organic-walled microfossils like cysts of certain heterotrophic dinoflagellates. *Lycopodium* tablets were added to enable the determination of absolute abundances. Slides were made using residues sieved at >10 microns for quantitative study of dinoflagellate cysts and pollen, and for the search for acritarchs.

3.2
Stable Isotopes

A total of 66 samples have been analyzed for stable oxygen and carbon isotope compositions, 47 from the Upper Pliocene interval (36 bulk rock samples, 11 dolomitized components and sparry cement subsamples), and 19 bulk rock samples from the Lower Pliocene interval. To obtain a matrix signal, larger components (> 0.2 mm) have been avoided for the bulk rock samples. Subsamples were removed with a dental drill device where larger components were present, or were taken from the ground bulk sample where larger components were absent. For the dolomitized components and sparry cement subsamples, a dental drill was employed.

Analyses were performed in the laboratory of M. Joachimski at the University of Erlangen/ Germany. Carbonate powder samples were reacted with 100% H_3PO_4 at 75°C in an online, automated carbonate reaction device (Kiel Device) connected to a Finnigan Mat 252 mass spectrometer. Isotopic ratios were corrected for ^{17}O contribution (Craig, 1957) and are reported in ppt relative to PDB (Peedee belemnite). External precision, based on multiple analyses of the NBS 19 standard and the CO 1 standard, is ±0.02‰ for $\delta^{13}C$ and ±0.04‰ for $\delta^{18}O$.

The data of the present study plot closely to the measurements of Melim et al. (in press-b). A slight systematic shift of the $\delta^{18}O$ composition towards lighter values is observed in the measurements carried out within the present study. This shift probably stems from the use of two different spectrometers.

3.3

X-ray Diffractometry

Carbonate mineralogy has been determined using the Phillips PW 1710 diffractometer of GEOMAR Research Center in Kiel/Germany. Bulk rock powder samples have been mounted on cavity mount holders. The samples were scanned from 25° to 40° to cover the significant peaks of the different carbonate minerals, at a scanning speed of 0.01° steps per second. A cobalt Kα tube was used at 40 kV and 40 mA.

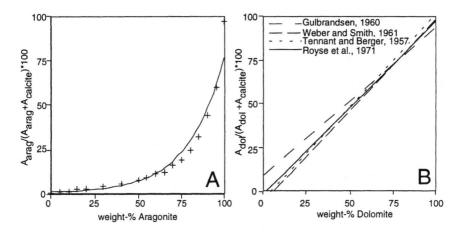

Fig. 10. Calibration curves for carbonate determination using X-ray diffraction. (**A**) Non-linear relationship between aragonite and calcite contents. (**B**) Roughly linear relationship between dolomite and calcite.

The percentage of the carbonate minerals (calcite, aragonite, dolomite) relative to total carbonate content was calculated from peak area ratios using calibration curves. These calculations are based on the observation that the relative amount of a mineral is related to the reflection intensity, which is proportional to the peak area in the diffractogram (Neumann, 1965). Peak areas were measured using a computer-based integration program. The non-linear relationship between calcite and aragonite (Milliman, 1974) was calculated from ratios calibrated from standard minerals measured on the same diffractometer at GEOMAR in Kiel/Germany (Fig. 10A). For the relative amounts of calcite and dolomite, a linear relationship was used based on Royse et al. (1971; Fig. 10B). Splitting of the calcite peak into a high Mg-calcite portion and a low Mg-calcite portion was found not to be required as no high Mg-calcite peaks were observed in the diffractograms.

Non-carbonate mineralogies like quartz, celestite, and clay were not quantified.

3.4

Scanning Electron Microscopy

To examine diagenetic alterations of the carbonate samples, a bulk rock preparation method was adopted similar to that of Lasemi and Sandberg (1984) as modified by Munnecke (1997). Bulk rock samples were cut perpendicular to the bedding, mounted on SEM stubs and polished with corundum powder 2000. Usage of polished surfaces was found to be superior to the examination of broken surfaces, because polished surfaces reveal more information on the internal structures of the components and of the cements, and are devoid of structures caused by breakage. Subsequently, they were cleaned and etched in 0.1 M hydrochloric acid for an optimum relief for SEM examinations and a better recognition of the boundaries of constituents. Also, dolomite and celestite are recognized more easily in etched samples by their elevated relief compared to calcite and aragonite. Etching time was standardized to exactly 20 seconds to obtain the same degree of etching in every sample. Organic-walled microfossils (palynomorphs) and other insoluble components, like terrigenous material, also become prominent during the etching procedure. Usually, larger palynomorphs are cut during polishing, allowing an observation of the vesicle filling and the thickness of the wall. Two uncemented samples (226.31 and 251.31 mbmp), however, have been investigated with unpolished, unetched surfaces in order to gain information on the preservation of aragonitic mud constituents. Each sample was finally coated with gold/palladium. Examinations were performed on a CamScan Scanning Electron Microscope Cs 44 of the Department of Geology and Paleontology at University of Kiel/Germany where 84 samples were investigated.

The examinations focused on a thorough description of diagenetic structures, especially the formation of mainly fine-grained cement. Also, diagenetic alterations of the sedimentary matrix and components were described. Grain sizes of cement crystals of the lithified samples have been analyzed by measuring the largest apparent diameter of these crystals on one to four micrographs per sample with a standard magnification of 1000x. Statistical analysis of the measured grain sizes of 46 samples were used to gain information on the influence of initial sediment composition on the style of cementation.

Stereoscopic pairs of SEM micrographs (with an angle of 6° between both micrographs) were employed to examine the three-dimensional shape of organic microfossils using a stereoscopic lens. This method is commonly applied in micropalaeontology and has recently been used for organic microfossils by Munnecke and Servais (1996).

Information on the chemical composition of constituents was gained by an EDX device integrated in the SEM. The EDX spectra was especially useful for the distinction between high Mg-calcite, low Mg-calcite, and dolomite. Also the identification of celestite and phosphate was performed using the EDX analyzer. The identification of aragonite and low Mg-calcite under the SEM is based on morphologic characteristics. As they show the same elementary composition they cannot be distinguished by straightforward EDX analysis. Identification of aragonite was performed using selective staining of aragonitic constituents with diluted Feigel's solution (Schneidermann and Sandberg, 1971). Aragonite, being slightly more soluble in acids, reacts faster with Feigel's solution. The solution leaves traces of Ag-precipitates on the aragonitic constituents. Using the EDX device, the stained aragonitic components are marked by small Ag-peaks, while calcitic constituents lack an Ag-peak.

3.5

Total Organic Carbon

Contents of organic carbon were determined with a CS Analyzer 125 (LECO) at GEOMAR in Kiel/Germany. 30 mg of a ground sample were treated with 0.25 M hydrochloric acid in order to remove the carbon fixed in the carbonate. The remainder of the samples were heated to 1200°C in an inductive oven. The organic matter is oxidized by this procedure, and the thus produced CO_2 and CO (transformed to CO_2 in a catalytic oven) is quantitatively determined by an infrared measuring device (Hölemann, 1994). To avoid inaccuracies, calibrations with a calcite standard were repeatedly carried out. Also, every sample was measured twice to minimize deviations. Measurements were rejected when they deviated by more than 0.03% of the weight of the sample prior to hydrochloride acid treatment.

3.6

Fluid Inclusions

Early sparry cements were studied for primary fluid inclusions. Microthermometrically measurable fluid inclusions were rare in the intervals examined, because sparry cements are only present in the few coarse-grained samples. Examinations of fluid inclusions were carried out by R. H. Goldstein, University of Kansas (Lawrence). For these measurements, doubly polished thin sections of 70 μm thickness were prepared. To prevent laboratory reequilibration of the fluid inclusions, the samples were prepared using cold techniques. Freezing runs were performed to determine the final ice melting temperatures of the fluid inclusions. Salinities were calculated from ice melting temperatures (Tm ice) using an NaCl equivalent curve:

$$\text{Salinity (wt\%)} = 1.78\ \Theta - 0.0422\ \Theta^2 + 0.000557\ \Theta^3$$

where θ is the depression of the freezing point Tm ice below 0°C. The NaCl equivalent curve has been used, as a reliable sea-water equivalent curve was not available for the high salinities found. For a detailed description of the method, the reader is referred to Goldstein and Reynolds (1994). The examinations were carried out on a Fluid Inc-Adapted USGS Gas Flow Heating and Freezing Stage at University of Kansas.

4 Primary Signals - Sediment Input

4.1

Periplatform Carbonates

Research on periplatform carbonates is intimately connected to the research on the Bahama Banks. Until the early 1970's, the Holocene platform top sediments were by far the best-studied part of the Bahamian geology. Sediment types, their constituents, and facies patterns were thoroughly described by Black (1933), Newell et al. (1951), Illing (1954), Newell and Rigby (1957), Cloud (1962), Purdy (1963), and others. A summary on depositional processes and environments is given by Bathurst (1971).

Description of the sedimentation in the deep-water realm commenced later and showed a progress that is coupled with advances of the drilling technology. In short gravity cores, Supko (1963) found high aragonite contents in core tops from Tongue of the Ocean and recognized the climatic significance of this observation. He suggested that this pattern reflects the flooding of the Bahama Banks during the latest sea-level rise, when aragonite needles, produced on the platform top, are winnowed into deeper water. For this type of sediment, the term "periplatform ooze" was coined by Schlager and James (1978). The distinct property of periplatform sediments is that they have two input sources, the shallow-water carbonate platform, producing material of mainly metastable mineralogy (aragonite and high Mg-calcite), and the open ocean, with material of mainly stable low Mg-calcite mineralogy.

Longer piston cores from Tongue of the Ocean and Exuma Sound made it possible to examine sediments reaching into the last glacial maximum (Pilkey and Rucker, 1966). Observed fluctuations in the mineralogy were interpreted to record glacial and interglacial periods. Fluctuations in sea-level could influence the mineralogy of the periplatform sediments by repetitively flooding and exposing the shallow-water carbonate factory. Aragonite-rich layers were thought to result from offbank transport of the fine-grained, predominantly aragonitic material produced within the platform interior during sea-level highstands (Pilkey and Rucker, 1966; Rucker, 1968; Kier and Pilkey, 1971; Lynts et al., 1973; Boardman and Neumann, 1984). Kier and Pilkey (1971) showed that not only the mineralogy but also the composition of the fine-grained periplatform ooze exhibits differences between aragonite-rich and calcite-rich layers. Aragonite-rich layers consist predominantly of platform-top derived aragonite needles, whereas calcite-rich layers contain large amounts of pelagic coccoliths. The fact that aragonite needles occurred in all samples (aragonite-rich as well as aragonite-poor ones) supported the assumption that the lower aragonite contents are an original input signal rather than caused by dissolution (Pilkey and Rucker, 1966; Kier and Pilkey, 1971). The observation that in the Tongue of the Ocean the sedimentation rates of interglacial periods exceed that of the glacials by a factor of four to six led to the model of highstand shedding: During sea-level highstands a productive carbonate platform exports aragonitic material, whereas during lowstands the exposed platform top ceases production (Droxler and Schlager, 1985).

Deep cores from ODP Leg 101 eventually allowed for the investigation of longer sedimentary records that revealed the existence of aragonite cycles continuing down into the Pliocene (Droxler et al., 1988). Similar to the fine-grained periplatform material, the sediments at the toe-of-slope of the Bahamas have been shown to be influenced by the differences in platform productivity during glacials and interglacials (Droxler et al., 1983; Cartwright, 1985; Reijmer et al., 1988). Droxler et al. (1983, 1988) showed that variations in the aragonite contents closely correlate to the stable oxygen isotope signal. High aragonite contents usually coincide with depleted oxygen isotope ratios. Although Droxler (1985) and Droxler et al. (1990) noted that off the Bahamas and the Maldives, the aragonite and oxygen isotope peaks can be slightly out of phase in either direction, the assumed climatic relevance of the aragonite signal was thus shown. The cause of the cyclic variations in the aragonite signal, however, are still subject to discussion (Reid et al., 1996). According to Boardman et al. (1986), the signal reflects exclusively variations in the input, whereas Droxler et al. (1983, 1991) ascribe the signal, at least in part, to climatically steered dissolution of aragonite by deep waters undersaturated in aragonite. Probably both mechanisms play a role, and the relative importance depends on the depth of deposition (Glaser and Droxler, 1993).

The slopes of the Bahama Banks themselves, however, have long received relatively little attention. Until the realization of the Bahamas Drilling Project, few studies have been carried out on the slope sediments. To mention are studies of Grammer and Ginsburg (1992) and Grammer et al. (1993-a) that are based on diving observations, and examined the present-day morphology and diagenesis of the slopes of Great Bahama Bank.

The significance of compositional variations with respect to sea-level variations is now well established. The restricted applicability of Holocene sedimentation as an actualistic model for the past of the Great Bahama Bank, however, was recognized by Schlager and Ginsburg (1981). Both, long-term natural evolution and outside factors were recognized to have caused significant changes in the morphology that resulted in a change in the sedimentary processes. Nevertheless, the effect of variations in the morphology of carbonate platforms, for example, is only sporadically considered in the literature (e.g. Burchette and

Wright, 1992; Bachmann et al., 1996; Bachmann and Willems, 1996). This chapter of the present study is dedicated to examine the effect of the morphological evolution as observed in the Pliocene of Great Bahama Bank. The questions that will be considered here can be summarized as follows:

(1) What is the expression of sea-level variations in the slope sediments penetrated by CLINO? (2) How can the compositional variations be explained in a genetic sense? (3) Are there differences in the compositional signature of sea-level variations in periplatform carbonates, with respect to the morphology of the carbonate platform?

4.2

Results of the Compositional Examinations

4.2.1

Components

Periplatform carbonates are typified by skeletal and non-skeletal constituents that show mixed characteristics of open marine, slope, and platform top environments. This typical composition is also observed in the Pliocene periplatform sediments from CLINO. The composition of the Upper and Lower Pliocene successions rather varies with respect to abundances than to occurrences. The shallow-water constituents found in both successions correspond to the biota found in time-equivalent shallow-water environments of the Great Bahama Bank (Beach, 1982; Kievman and Ginsburg, in press).

On the modern Bahamas, non-skeletal constituents predominate within the shallow-water environment (< 20 m; Fig. 11). Grain types from the Holocene of the Bahamas have been described by Illing (1954), Newell and Rigby (1957), Purdy (1963), Ball (1967), Milliman (1974), Bathurst (1975), and Palmer (1979). Pellet deposition has been described specifically by Cloud (1962) and Purdy (1963). The composition of periplatform sediments, however, is not a straight-forward mirror of platform top sedimentation, as usually fine-grained material is exported to the slopes while coarser grains are left behind on the platform top (Fig. 11). Therefore, platform top sediments are usually coarser-grained than the corresponding slope sediments (Freile et al. 1995).

The components described below have been examined in thin sections under the light microscope, and subordinately under the SEM. Palynomorphs are described from light microscope and SEM examinations. The distribution of the different components throughout the intervals of CLINO is given in Chapter 4.2.2. Skeletal components are often considerably altered by diagenetic processes (see Chapter 5). Therefore, a typification of the biota down to species level is renounced. Often it could not be decided if only one or several species of a genus were present. Consequently the typification using "sp." and "spp." is not applied. A determination to species level was performed for the well-preserved palynomorphs by M. Head (University of Toronto).

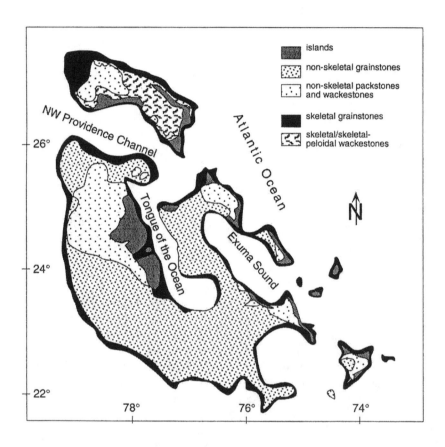

Fig. 11. Surface sediment composition on the present-day Bahama Banks. Surface sediments on Great Bahama Bank are characterized by a predominance of non-skeletal grains. Most parts of the submerged platform top of Great Bahama Bank are covered by grainstones, i.e., the amount of fine-grained sedimentary matrix is low. Over large areas, the fine-grained material is flushed out by water movement. Fine-grained material is present mainly in the pack- to wackestones of the leeward "shadows" of the islands. Skeletal grainstones occur at the margins of the bank. From Enos (1974), modified with the permission of the publisher, the Geological Society of America, Boulder, Colorado USA. Copyright©1974 Geological Society of America.

4.2.1.1
Skeletal Components

Flora. The flora is characterized by larger sessile and small pelagic algae, and different sporomorphs.

<u>*Algae.*</u> Green (Chlorophyta) and red algae (Rhodophyta) are common in the Pliocene periplatform carbonates of CLINO. Chlorophyta are represented by the families Codiaceae and Dasycladaceae, Rhodophyta by the family Corallinaceae.

Chlorophyta. Codiacean algae are thought to belong to the most prolific sediment producers in shallow warm waters (Ginsburg, 1956; Stockman et al., 1967; Milliman, 1974; Neumann and Land, 1975; Wray, 1977). *Halimeda* (Lamouroux, 1812; Plate 1A) is the most common calcifying codiacean algae observed in the Pliocene of CLINO, being a dominant constituent in some coarse-grained samples, whereas it is present in lower amounts in many other samples. Various sizes of fragments (up to 10 mm in length, usually smaller) are observed. The plates with irregular tubes usually appear brownish under the light microscope. Neomorphosed specimens show a characteristic pseudopleochroism (see Chapter 5). *Halimeda* is described to grow abundantly in shallow-water, turbulent environments, but also occurs commonly on the upper slope (Ginsburg et al., 1991; Freile et al., 1995; Dullo et al., in press).

Other Codiacean algae could not be identified. Nevertheless, as they are present on the modern Great Bahama Bank (Neuman and Land, 1975), *Penicillus* (Lamarck, 1813), *Rhipocephalus* (Kuetzing, 1843), and *Udotea* (Lamouroux, 1812) have presumably also grown on the Great Bahama Bank in the Pliocene. Post-mortal break down of such green algae results in mud-sized aragonite needles with indistinguishable origin (e.g. Lowenstam and Epstein, 1957; Matthews, 1966; Stockman et al., 1967). The contribution of Codiaceans to the aragonite needle sediment budget is controversially debated (e.g. Neumann and Land, 1975; Loreau, 1982; Macintyre and Reid, 1992; see below in description of fine-grained matrix).

Dasycladacean algae are rare in the samples examined. Only few small fragments were identified. This corresponds to the rarity of Dasycladacean remains in the shallow subsurface of Great Bahama Bank and Little Bahama Bank as described by Beach (1982) and Williams (1985), respectively. Specimens found in platform top sediments of the same age have been described by Beach (1982) as *Cymopolia* (Lamouroux, 1816).

Rhodophyta. Branching as well as encrusting corallinaceans are present. Encrusting forms, represented by Melobesioidae (Bizzozero, 1885; Plate 1B), are rare and restricted to an interval in the Upper Pliocene succession (256.2 - 257.4 mbmp). They form nodules of up to 12 mm in diameter that are characterized by an irregular concentric-laminated growth pattern around a variety of substrates like e.g. *Halimeda* fragments and bryozoans. However, details of the growths patterns as, for example, breakage of branches due to high water energy, could not be observed. This is partly because of the small size of most fragments, and partly because the fine growth structures as described in the literature (Bosence, 1977; Dullo, 1983) are obscured by dolomitization. Nevertheless, the globular shape of the nodules indicates a high-energy source area.

Fragments of branching coralline algae, belonging to the Corallineae (*Corallina*; Linnaeus, 1758; Plate 1C), are sporadically found throughout the selected intervals, but never exceed 3 % of the total sediment. Their characteristic fine-cellular structure is preserved even in strongly dolomitized fragments. Frequently, the central hypothallus and the denser perithallus can be distinguished.

Modern red algae precipitate carbonate down to depths of 200 m below sea-level (Adey and Macintyre, 1973). Among the deep-water forms are mainly slowly growing rhodoliths (Bosellini and Ginsburg, 1971). Most forms, however, grow in water depths shallower than 50 m (Milliman, 1977, and citations therein), unless extremely clear water, as present in the Red Sea, allows for a deeper light penetration (Dullo et al., 1990).

Chrysophyta. Coccolithophoridae are rare constituents of the fine-grained matrix. They are estimated to account for less than 1% in the matrix of some SEM samples, but usually are absent. Usually, single coccoliths are found (*?Umbilicosphaera sp.*; Plate 1D), whereas undisaggregated coccospheres are extremely rare. Scarce epitaxial cement overgrowths are observed.

Pyrrhophyta. Dinoflagellate cysts are common accessory constituents. Preservation varies from fair to excellent. The assemblages (as determined by M. Head, Toronto) are characterized by a low diversity (Appendix 2; Plate 2). *Polysphaeridium zoharyi* (Rossignol, 1962; Plate 2Q) is common especially in the Lower Pliocene samples. It is an inner neritic, tropical to subtropical species adapted to high salinities and has been described from the present-day Bahamas and the Persian Gulf (Morzadec-Kerfourn, 1983, 1992; Bradford and Wall, 1984). At present, *P. zoharyi* only occurs abundantly where summer sea surface temperatures are greater than about 28°C (Edwards and Andrle, 1992). In the Upper and the Lower Pliocene samples of the present study, the thermophil *Lingulodinium machaerophorum* (Deflandre and Cookson, 1955; Plate 2J) also is common. As it prefers euryhaline conditions, it is negatively correlated to *P. zoharyi* . In modern sediments of the North Atlantic, *L. machaerophorum* is most abundant where winter and summer sea-surface temperatures are above 15°C and 27°C, respectively (Edwards and Andrle, 1992). *Hystrichokolpoma rigaudiae* (Deflandre and Cookson,

1955; Plate 2G) and *Melitasphaeridium choanophorum* (Deflandre and Cookson, 1955; Plate 2K) also are clearly thermophil (Head, 1997).

Spiniferites spp. (Mantell, 1850; Plate 2O), a cosmopolitan neritic group, is abundant. Moderate numbers of the subtropical to tropical genus *Operculodinium* (Wall, 1967; Plate 2M and N) were found, including *O. israelanum* (Rossignol, 1962), and the two new species defined within the course of the present study, "*O. bahamense*" and "*O.? megagranum*" (Head in Head et al., in prep.). Rare specimens of *Tuberculodinium vancampoae* (Rossignol, 1962; Plate 2R), that are abundant in modern low salinity estuaries points to a mixed origin of the assemblage (Wall et al., 1973, 1977). *Selenopemphix quanta* (Bradford, 1975; Plate 2P) are also present that show highest abundances in shallow, tropical to subtropical waters of less than 40 m depth (Edwards and Andrle, 1992). Neritic species such as *Dapsilidinium pseudocolligerum* (Stover, 1977; Plate 2F), *Nematosphaeropsis rigida* (Wrenn, 1988; Plate 2L), *Impagidinum paradoxum* (Wall, 1967; Plate 2H), and *I. plicatum* (Versteegh and Zevenboom, 1995), that could indicate an open marine influence, are rare.

Outer shelf and oceanic species appear to be absent. The assemblages compare closely with modern assemblages off the west coast of Andros Island (Wall et al., 1977) with significant numbers of tropical species of which several are typical for elevated salinities. In particular, *Polysphaeridium zoharyi* is a tropical species adapted to saline conditions (Morzadec-Kerfourn, 1983, 1992) as occur today on the Bahama Bank.

Acritarchs. Acritarchs are found infrequently in SEM bulk rock samples, and in palynologic slides (Appendix 2). Due to their small size (< 5 μm) they are presumably often overlooked. The taxonomy and paleoecology of Cenozoic acritarchs is very poorly known. The specimens found (of which most are apparently undescribed; M. Head, written communication, 1997) might be of marine origin but could also be freshwater algal cysts washed into the marine realm. Some might be assignable to the acritarch genus *Cymatiosphaera* .

Land plants. Determinable remains of land plants are restricted to rare trilete fern spores, bisaccate pollen of *Pinus*, and rare angiosperm pollen that could not be determined to species level (Plate 2C, D, E). *Pinus* pollen are characterized by a central body with two bladders (sacci) on either side. They are dispersed by wind and, once dropped in the ocean, behave hydrodynamically similar to mud particles.

Fauna. Organisms that originate within different sedimentary facies belts along the platform-slope transect are represented mainly by benthic and planktic foraminifers, and by a variety of sessile organisms that characterize reefal and other sedimentary realms.

Scleractinia. A large number of species have been described from platform top and upper slope deposits that are approximately time-equivalent to the intervals investigated in this study (Beach and Ginsburg, 1980; Kievman and Ginsburg, in press). Nevertheless, no remains of scleractinians were identified in the two Pliocene periplatform successions of CLINO examined in this study. The reason for the absence of determinable coral fragments can only be speculated. The aragonitic skeleton might have been altered too strongly to be recognized, or might have been dissolved leaving non-characteristic molds. Constantz (1986) and Dullo (1986) have demonstrated the high susceptibility of corals to diagenetic alterations that is caused by the small size of the aragonitic skeletal crystallites. Therefore, a large portion of the non-determinable grains (Plate 4E; see below) that are particularly abundant in the selected Lower Pliocene interval possibly originate from coral skeletons.

Mollusks. Gastropods frequently occur in the periplatform sediments. Leaching and recrystallisation of the primary aragonitic shells hampered further identification of most fragments, as the original wall structure is usually lost (Plate 1E). Sometimes slight pleochroism indicates that some submicroscopic ghost structures are preserved. Bivalve fragments are rare, and preservation varies (Plate 1F). Pteropods are extremely rare. Their primarily aragonitic test has a relatively low potential for fossilization (Plate 7G).

Bryozoa. The rare bryozoans present belong to the order Cheilostomata (Busk, 1852). Encrusting bryozoan colonies of the family Schizoporellidae (Jullien, 1903) (possibly *Schizoporella*; Hincks, 1877; Plate 1G) with irregular chambers have been observed in some samples. In some coarse-grained samples, specimens of *Adeona* (Lamouroux, 1812; Plate 1F) could be identified. They reach sizes of up to 2 mm in length, and are characterized by their regular bisymmetrical shape.

Foraminifera. The variety of foraminifers present in the slope sediments examined is characteristic for an environment with a mixed input from shallow-marine, deeper-marine, and open-marine sources. Benthic and planktic foraminifers from the Bahamas have been previously described by, e.g., Illing (1950, 1952), Rose and Lidz (1977), Lidz and Bralower (1994), and Lidz and McNeill (1995-a).

Benthic foraminifers are a common constituent throughout most of the Pliocene periplatform sediments. Most genera identified belong to the families Miliolidae, Soritidae, Amphisteginidae, Discorbidae, Rotaliidae, Homotremidae, and Textulariidae. Benthic foraminifera tests, mostly composed of calcite with varying amounts of $MgCO_3$ are generally moderately to well preserved. Small Miliolidae, however, with thin, microcrystalline high Mg-calcitic tests are sometimes preserved as molds. In palynologic slides, foraminiferal linings are frequently observed (Plate 2B). Planktic foraminifers are also frequent throughout the intervals examined. They show a greater relative and absolute abundance in the Lower Pliocene succession than in the Upper Pliocene interval.

The classification used in this study is based on Loeblich and Tappan (1964), Lidz and Bralower (1994), and Lidz and McNeill (1995-a). Environmental interpretation of the groups distinguished follows Murray (1973), Rose and Lidz (1977), and Lidz and McNeill (1995-a).

Milliolacea (Ehrenberg, 1839): Most of the Miliolaceans belong to the Miliolidae (Ehrenberg, 1839). Small rotund forms (Plate 3A; e.g. *Miliolinella*; Wiesner, 1931, *Triloculina*; d'Orbigny, 1826) occur in vast amounts in some fine-grained packstones. Small specimens of *Quinqueloculina* (d'Orbigny, 1826) are also present. These small forms have thin tests which are sometimes dissolved, leaving tiny molds. Larger forms like *Pyrgo* (Defrance, 1824) are rare (Plate 3B). Wall structures are usually largely obliterated.

According to Rose and Lidz (1977) and Murray (1973), Miliolidae are interpreted as indicators of shallow-water input. They are abundant in platform top sediments of the modern Bahamas, especially in water depths shallower than 3 m. In extremely shallow-water, rotund forms predominate. They are found to prefer mudbank habitats (Rose and Lidz, 1977). *Triloculina* and *Quinqueloculina* are found in hypersaline lagoons (i.e. elevated saline sea-water lagoons), and *Miliolinella* is tolerant to elevated salinities of up to as much as 50‰ (Murray, 1973). *Pyrgo*, in contrast, prefers normal marine environments.

Rare specimens of *Spiroloculina* (d'Orbigny, 1826) represent the family Nubecullariidae (Jones, 1875). Soritidae (Ehrenberg, 1839) are represented by rare large, multichambered forms of *Sorites* (Ehrenberg, 1839) and infrequent *Peneroplis* (de Montfort, 1808). These foraminifers, that are symbiotic with Zooxanthellae, are characteristic for shallow-water environments such as sea-grass areas. They are resistant to elevated salinities (Murray, 1973).

Textulariina (Delage and Hérouard, 1896): The specimens of this group, that are characterized by their agglutinating wall, sporadically occur (Plate 3C). Textulariidae (Ehrenberg, 1838), represented by *Bigenerina* (d'Orbigny 1826), and Ataxophragmidae (Schwager, 1877) cf. *Karreriella* (Cushman, 1933) with characteristic slightly twisted tests, are present. They have been described to typify cooler water environments in depths greater than 100 m (Murray, 1973).

Rotaliina (Delage and Hérouard, 1896): Rotaliidae (Ehrenberg, 1839) are abundant (e.g. *Rotalia*; Lamarck, 1804). Their tests are recognized by a transparent appearance under the light microscope (Plate 3D). In non-oriented thin sections, further taxomomic typification is difficult. Many Rotaliidae live in sediment-covered, shallow-water areas (Murray, 1973).

Discorbidae (Ehrenberg, 1938), represented by *Rosalina* (d'Orbigny, 1826) is abundant in some fine-grained packstones (Plate 3E). *Rosalina* is characterized by a concave umbilical side, and frequent preservation of the inner organic layer. On the recent Bahamas, *Rosalina* is found in outer shelf sediments and is closely related to the presence of reefs (Rose and Lidz, 1977). *Rosalina* is frequently found clinging to vegetation, or hard substrates in normal marine waters (Murray, 1973).

Amphisteginidae (Cushman, 1926) are represented by *Amphistegina* (d'Orbigny, 1926) that is easily discernible by its characteristic large size (up to 2 mm in long-axial section) and the swollen, massive test (Plate 3F). Amphisteginids have a limited distribution and flourish in shallow-water settings such as outer-margin- and shallow-slope environments. On the modern Bahamas they show a marked preference for coral reefs along exposed windward edges (Rose and Lidz, 1977), but also occur in sea-grass areas (Murray, 1973).

Cibicididae (Cushman, 1927), namely *Cibicides* (de Montfort, 1808) is infrequently observed. The radial microtexture makes *Cibicides* easily recognizable. *Cibicides* occurs in a variety of water depths and temperatures and therefore is not diagnostic. Buliminidae (Jones, 1875) are rare. *Bulimina* (d'Orbigny, 1826) and *Reussella* (Galloway, 1933) are indicative of deeper-water conditions. *Reussella* is easily discernible by its triserial, angular shape. Elphidiidae (Galloway, 1933), typical for reefal environments, and Eponididae (Hofker, 1951) occur in low amounts. Homotrematidae (Cushman, 1927) are sporadically observed as encrusters of bioclastic grains. They grow preferably in reefal areas (Murray, 1973).

Globigerinacea (Carpenter, Parker and Jones, 1862) are present throughout the selected Pliocene intervals. The joint occurrence of planktic with benthic genera is characteristic for periplatform carbonates. Most planktic specimens observed belong to the subfamilies Globigerininae (Carpenter, Parker and Jones, 1862; e.g. *Globigerinoides*; Cushman, 1927; Plate 3G) and Sphaeroidinellinae (Banner and Blow, 1959; e.g. *Spheroidinella*; Cushman, 1927). Less abundant are Orbulininae (Schultze, 1854; *Orbulina*; d'Orbigny, 1839; Plate 3H). Rarely Globorotaliidae (Cushman, 1927) occur (*Globorotalia*; Cushman, 1927).

Arthropoda. Thin, smooth shells of small Ostracoda are ubiquitous (Plate 4A). Mostly, ostracods are preserved as single valves or fragments, two-valved specimens are rare. Sweeping extinction under crossed Nicols results from the homogeneous prismatic wall structure. These benthic crustaceans are adapted to a wide variety of environments. Fragments of Cirripedia (barnacles) are extremely rare. Their tests are distinguished by subparallel darker lines. Barnacles occur in a wide range of water depths.

Echinodermata. Echinoderms are only minor constituents in all samples (Plate 4B). Many echinoderm fragments are easily identified by their characteristic single-crystal extinction pattern, that is frequently intensified by epitaxial cement overgrowth. The uniform sieve-like microtexture is sometimes obliterated by dolomitization. Echinoid spines most likely originate from Echinidae (Gray, 1825), Ophiuroidae (Lyman, 1865), or Asteriidae (Gray, 1840). Further determination that could lead to an exact environmental interpretation is frequently hindered by the small size of the fragments present.

Annelids. Calcareous tubes of the Polychaete worms (family Serpulidae) with concentric laminae of calcite encrusting skeletal debris are rarely observed. Serpulids, being typical pioneer biota are soon replaced by other more competitive biota, and therefore are usually rare in the fossil record (Schuhmacher, 1977; Goren, 1979; Vine and Bailey-Brock, 1984). Mandible specimens observed in palynologic slides are probably the mouthparts of annelid worms (i.e. scolecodonts; Plate 2A).

Vertebrate remains. Possible remains of vertebrates are the scarce spherical massive components that appear brownish under light microscope. They are interpreted as otolites.

Boring organisms and burrowers. biogenic components are frequently altered by boring organisms. Beside the micrite rims of cortoids (see below), small tubes are common that cut the structure of e.g. Rotaliid tests (Plate 4C). Algae are thought to have produced the borings, but bacteria and fungi are also possible borers that could have formed the structures found (compare Zeff and Perkins, 1979). The borings, if they are of algal origin, could indicate that the grain was derived from the photic zone (Scholle, 1979).
Burrowing organisms left their traces in fecal pellets (Plate 4D). Under the SEM, aragonite needles that are the main constituents of the pellets, appear aligned to the burrows, which is characteristic for soft-material coprophags (K. P. Vogel, written communication, 1996). These burrows are not refilled by cements.

Non-determinable grains. Numerous grains could not be identified either because of diagenetic alterations (recrystallization or dissolution) that obscured the microstructures of the biodetrital grains, or because of their small size that made determination of specific characteristics impossible (compare Dullo, 1983, 1984; Plate 4E). Recrystallization as well as micritization are observed. Molds are also common, especially in the Lower Pliocene succession. Of the cryptocrystalline grains, neomorphous grains, and molds, a large fraction is thought to represent altered skeletal grains of metastable primary mineralogy (aragonite and high Mg-calcite). Remains of scleractinian corals could possibly make up considerable amounts of the non-determinable grains, because their skeletons are highly susceptible to diagenesis (Constantz, 1986; Dullo, 1986).

4.2.1.2
Non-Skeletal Components

Cortoids. Cortoids (Flügel, 1982; Plate 4F), defined by a micritized rim (Bathurst, 1971), are usually composed of rounded bioclasts. The micrite rim is thought to originate from borings of micro-organisms such as different algae and fungi (Bathurst, 1966). Frequently the original grains are leached leaving the emptied or internally cement-filled micrite envelopes. Large amounts of cortoids are observed that have formed of *Halimeda* debris.

Micrite envelopes, formed by boring microorganisms, are considered typical for shallow-water areas (< 20 m; e.g. Swinchatt, 1969; Gunatilaka, 1976; Flügel, 1982). They are thought to form in warm-water supersaturated with respect to calcium carbonate, because the micrite that infills the borings is thought to be a cement. Although micritic rims are also described from deeper water that is saturated with respect to carbonate (Zeff and Perkins, 1979; Hook et al., 1984), they are significantly more abundant in shallow marine environments (Bathurst, 1966).

Intraclasts. Intraclasts are locally-derived, penecontemporaneously eroded fragments of partly consolidated sediment (Folk, 1959). In the samples examined, intraclasts usually consist of a dominantly micritic material. They are characterized by rounded shapes and incorporated fine grains which are indented rather than cut by the boundaries of the intraclasts. This indicates that the material was still soft at the time of reworking (Plate 4G). Intraclasts are rare and occur only in coarse-grained deposits.

Fine-grained matrix. As observed with the SEM, aragonite needles (around 5μm in length) are abundant matrix constituents, especially in the upper interval examined (Plate 4H). The skeletal versus non-skeletal origin of aragonite needles has been controversially discussed. While e.g. Lowenstam and Epstein (1957) and Neumann and Land (1975) propose the aragonite needles to originate from the decay of codiacean algae, Cloud (1962), Steinen et al. (1988), and Shinn et al. (1989) consider precipitation from oversaturated sea-water. The morphological differences between skeletal and non-skeletal aragonite needles observed by Macintyre and Reid (1992), however, did not aid identification of the origin of aragonite needles in the present study, as slight dissolution/precipitation features have altered the shape of the needles. The poorly developed crystal faces of the needles in CLINO might point to an inorganic origin from whitings (compare Macintyre and Reid, 1992). Loreau (1982) has shown that only 25-40% of the crystals in many calcareous algae are aragonite needles, but most crystals are equant grains.

The fine-grained matrix especially in the Upper Pliocene succession, however, is dominated by aragonite needles, while equant grains are subordinate. In the Lower Pliocene succession, aragonite needles are less abundant. Here, calcite crystallites of uncertain origin dominate the fine-grained matrix. It is assumed that the majority of these calcite crystallites are of biodetrital origin. Aragonite crystallites of various shapes could have precipitated as epicellular biominerals on picoplankton blue-green algae and other organic substrates (Robbins and Blackwelder, 1992).

As will be shown in Chapter 5, in samples from both, the Upper and the Lower Pliocene selected intervals, aragonite needles and other mud-sized matrix constituents become enclosed in microspar-sized cement crystals. Microspar thus represents lithified sedimentary mud. Therefore, in quantitative component analysis (Chapter 4.2.2), both micrite (i.e. sedimentary mud) and microspar (being a product of diagenesis) are treated together as fine-grained matrix.

Microdolomite rhombs of diagenetic origin represent subordinate fractions of the fine-grained material throughout the intervals examined.

Peloids. Peloids are the most abundant non-skeletal grains in the intervals studied. They are the dominant grains in some microfacies (see Chapter 4.2.5.). Peloids are oval to rounded and range in length from 0.1 to 1.5 mm. They exhibit a fine-grained internal texture under the light microscope. The interior of peloids is occasionally dissolved. SEM examinations revealed that fecal pellets are composed almost exclusively of aragonite needles. EDX analyses after staining with Feigel's solution (method after Schneidermann and Sandberg, 1971) has proven the aragonitic mineralogy of the needles.

Peloids are multigenetic. By far the most common type are the fecal pellets described before, as is indicated by tiny burrows in their margins that are typical for soft sediment burrowers (Plate 4D). In strongly recrystallized samples, however, cryptocrystalline grains (included in point-count group of non-determinable grains) could sometimes not be unequivocally distinguished from fecal pellets. Also, strongly compacted peloids in some samples make up a matrix-like fabric that is difficult to distinguish from a sedimentary matrix. Peloidal internal cements like those described by Reid et al. (1990) also occur (Chapter 5; Plate 11A). They are of purely diagenetic origin and are not considered as primary components.

On the platform top of the recent Great Bahama Bank, peloids constitute a large fraction of the sediment (Enos, 1974). Thus they are thought to be strong indicators for platform top productivity also in the Pliocene deposits.

Ooids. Ooids are absent from the Pliocene interval examined in the present study. Pliocene platform top sediments have been shown to contain only low amounts of ooids, in contrast to the overlying Pleistocene upper Lucayan Formation where ooids are found in large amounts (Beach and Ginsburg, 1980). Similarly, in pre-

Pleistocene turbidites at the toe-of-slope of Great Bahama Bank, that are time-equivalent to the intervals examined here, ooids are rare (Reijmer et al., 1992). The onset of considerable ooid deposition in the Pleistocene on the platform top as well as at the toe-of-slope is interpreted to be associated with progressively shallowing water on the banktop (Beach and Ginsburg, 1980; Reijmer et al., 1992).

4.2.2
Quantitative Compositional Analysis

Point counting led to the recognition of distinct patterns of the 14 groups of components that have been distinguished: (1) *Halimeda*, (2) red algae, (3) mollusks, (4) bryozoans, (5) benthic foraminifers, (6) planktic foraminifers, (7) ostracods, (8) echinoderms, (9) non-determinable grains, (10) cortoids, (11) intraclasts, (12) fine-grained matrix, (13) peloids, and (14) primary interparticle pore space (generally preserved as sparry cement). It will be shown that the selected Upper Pliocene succession exhibits a wider variability of the relative abundances and reveals clearer trends than the selected Lower Pliocene interval.

4.2.2.1
Distribution within Upper Pliocene Interval

The selected Upper Pliocene interval in foraminiferal zone N22 serves as an example of periplatform sediments of a steep-sided, flat-topped carbonate platform. It exhibits two distinct types of sediments: thick fine-grained intervals, and thin coarse-grained sections (Fig. 12). They correspond to the macroscopically distinguishable alternations of grain- to packstones, with the more uniform wackestones. The gamma-ray log shows high values in the lower coarse-grained interval (around 256 mbmp), while the upper coarse-grained interval is reflected by only a slight increase in gamma-ray values (around 222 mbmp; Figs. 7 and 12).

The fine-grained intervals (217.0 - 219.5 mbmp, 256.0 - 231.5 mbmp) are characterized by a predominance of sedimentary matrix averaging around 70% and in some samples exceeding 90%. Cortoids and intraclast, characteristic for rim environments, are virtually absent from these fine-grained samples. Mollusks, echinoderms, and ostracods occur subordinately. These biota are non-diagnostic with respect to distinct facies belts. Fine-grained biodetritus and other non-determinable grains account for about 15 %. Planktic foraminifers are frequently present. Benthic foraminifers increase towards the top of the fine-grained intervals, showing maxima around the base of the overlying coarse-grained deposits. Parallel to the benthic foraminifers, peloids gradually increase upcore towards the top of the fine-grained intervals, resulting in a coarsening-upward trend. Their maxima, exceeding 40% of the constituents, are located slightly below that of the benthic foraminifers. In some strongly compacted samples (e.g. 261.7 - 261.9 mbmp), the true amount of peloids might be higher as homogenized peloids can appear as a pseudomatrix. Both, benthic foraminifers and peloids exhibit a subordinate increase in abundance around 244 mbmp, that is also mirrored in the gamma-ray log.

Above the peloidal packstones at the top of the fine-grained intervals, an increase in cortoids and intraclasts marks a drastic change in sediment composition (at about 261.5 and 224.0 mbmp). Cortoids, being virtually absent from the fine-grained layers, suddenly exceed amounts of 50%. Parallel to this trend, the amount of sparry cement increases, i.e. the amount of sedimentary matrix decreases. Besides cortoids, the coarse-grained intervals are mainly composed of intraclasts, benthic foraminifers, and fragments of red and green algae. These intervals are thus mainly composed of biota that are characteristic of rim to upper slope facies belts. Bryozoans occur exclusively in the coarse-grained intervals. The top of the coarse-grained layers is defined by an abrupt decrease in cortoids and an equally abrupt increase in fine-grained matrix.

4.2.2.2
Distribution within Lower Pliocene Interval

The Lower Pliocene interval from biozone N19 represents slope sediments from a distally-steepened ramp. The point-count results reveal a rather uniform composition (Fig. 13). The fine-grained matrix averages around 50% throughout the entire succession. Sparry cement is absent. A large fraction of the components distinguished consists of non-determinable fine-grained biodetritus. Peloids show maximum amounts of 15%. Three subtle cycles in the abundance of peloids are observed that show a rapid increase in quantity at their bases (501.50, 491.50, and 483.50 mbmp) and a more gradual decrease upwards. The benthic foraminifers may parallel the peloids in the upcore trend by increases in abundance at 502.50, 491.50, and 481.50 mbmp. The saw-tooth shape of the gamma-ray curve with peaks at 477 and 498 mbmp is not found again in the cyclicities of benthic foraminifers or peloids. Bryozoans are extremely rare. In analogy to the Upper Pliocene interval, their occurrence at 576.5 mbmp could indicate platform rim input. Planktic foraminifers generally are more abundant than in the N22 succession. A strong peak with up to 20% of planktic foraminifers characterizes the base of the interval examined. A similar maximum at the base of the interval examined is exhibited by the echinoderms.

Components that in the Upper Pliocene show a characteristic distribution and are indicative for rim and upper slope environments such as cortoids, intraclasts, *Halimeda*, and red algae are virtually absent.

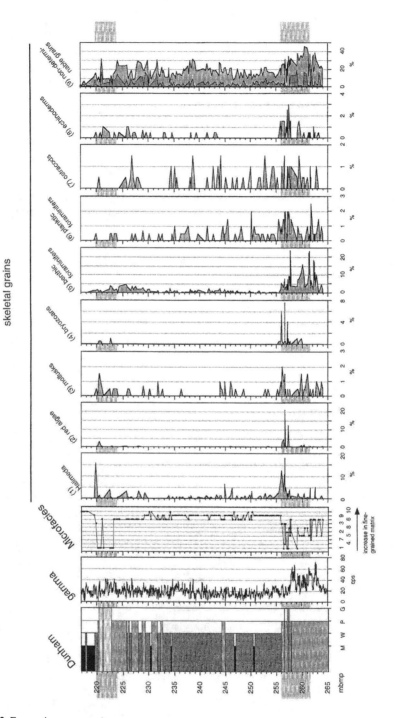

Fig. 12. For caption see opposite page.

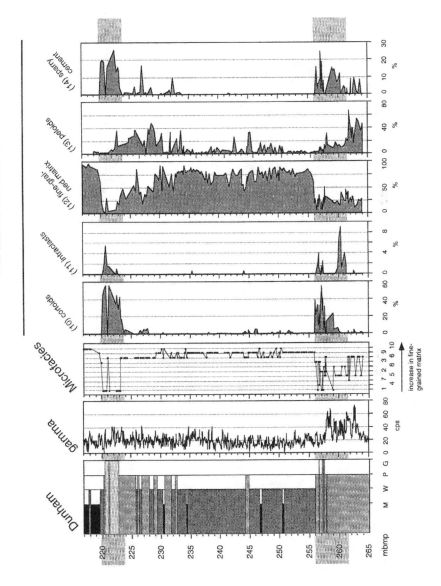

Fig. 12 contd. Results of components analysis of the selected Upper Pliocene interval. The abundance of the different skeletal and non-skeletal components is shown versus depth. Darker grey curve overlain on the (9) non-determinable grains shows the fraction of molds that contribute to this group. Note the strong variations in the percentage of fine-grained matrix, and the antithetic variations in the amounts of cortoids and sparry cement. Coarse-grained intervals are underlain by a grey pattern. The left column shows the classification after Dunham (G=grainstones, P=packstones, W=wackestones, M=mudstones). The second column shows the gamma-ray curve of Kenter et al. (in press), and the third column shows the microfacies (described below in Chapter 4.2.5).

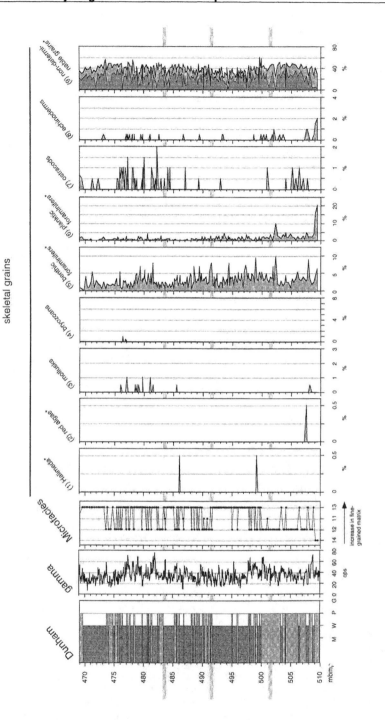

Fig. 13. For caption see opposite page.

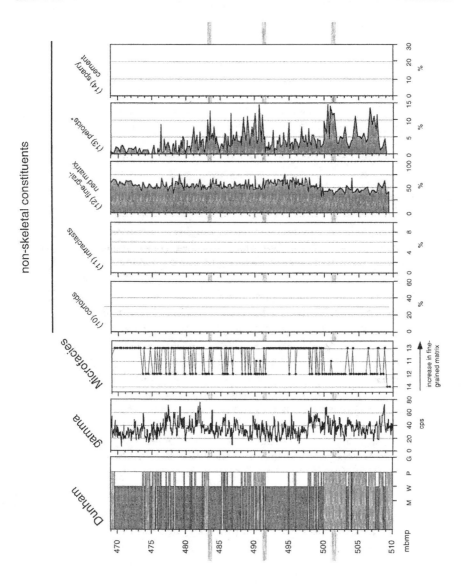

Fig. 13 contd. Results of components analysis of the selected Lower Pliocene interval. The densely sampled interval between 469 and 510 mbmp is shown. Darker grey curve overlain on the non-determinable grains shows the fraction of molds that contribute to this group. Variations in the composition are less pronounced than in the Upper Pliocene succession. Three subtle cycles are observed in the abundances of benthic foraminifers and peloids. Their bases are marked by grey lines. X-axis scale is the same as in figure 13 unless the component is marked by an asterisk. The left column shows the classification after Dunham (G=grainstones, P=packstones, W=wackestones, M=mudstones). The second column show the gamma-ray curve of Kenter et al. (in press), and the third column shows the microfacies (described below in Chapter 4.2.5).

4.2.3
Numerical Analyses of the Point-Count Results

Statistical methods were applied to estimate the behavior, significance, and inter-relationships between the different groups of components that have been distinguished in point-counting.

4.2.3.1
Statistics of the Upper Pliocene Composition

Numerical analysis of the data set of 155 roughly equidistant Upper Pliocene samples led to the following observations. The frequency distribution of specific point-count groups from the selected interval in foraminiferal zone N22 shows specific patterns for the different point-count groups (Fig. 14). The histogram of the fine-grained matrix is characterized by a bimodal distribution. One peak plots around 25%, the other around 85% of the total sediment volume. This indicates that there is a fundamental difference between the two types of deposits. Other histograms (e.g. those for cortoids and sparry cement) exhibit an asymptotic shape (high skewness), reflecting that in most samples these point-count groups are rare to absent but in some samples they are common to abundant (Table 1).

For most point-count groups, the skewness has values greater than 1.5, indicating a strongly asymmetric distribution. The category of non-determinable grains exhibits a normal distribution with a lower skewness of 0.67. This group is not a genetic one and partly depends on factors (mainly diagenesis) that differ from those of the other point-count groups that directly reflect sediment input.

The sensitivity of a point-count group, with respect to changes in the input, is expressed by its coefficient of variation, that is the standard deviation over the mean. Point-count groups which are spatially clearly restricted to a definite facies belt are characterized by high variation coefficients (Table 1). In the selected succession of zone N22, the highest coefficients are found for red algae, bryozoans, intraclasts, *Halimeda*, and cortoids in descending order. For most of these groups an origin from rather well defined facies zones is assumed, like in the case of the red algae the high energy fore slope.

Correlation coefficients characterize the statistical dependence between the different point-count groups. The highest positive correlation is observed between sparry cement and cortoids (0.80; Table 2). Cortoids also correlate positively with red algae and bryozoans (0.36 and 0.37, resp.), that are thought to originate from a similar high energy environment. As expected, cortoids as well as sparry cement (primary interparticle space) correlate strongly negatively with the fine-grained matrix (-0.64 and -0.72, respectively). The negative correlation between fine-grained matrix and peloids (-0.61) and between fine-grained matrix and benthic

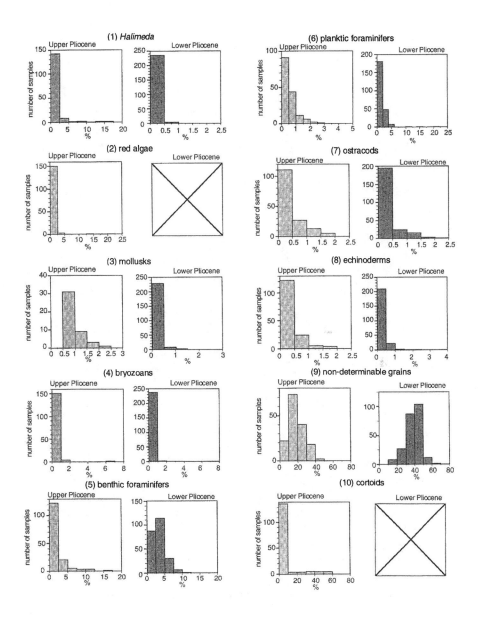

Fig. 14. Frequency distribution of constituents from the selected Upper (left) and Lower Pliocene intervals (right). Samples are shown that are included in statistical analyses (Upper Pliocene: n=155; Lower Pliocene: n=225). Note the bimodal distribution of (12) fine-grained matrix in the Upper Pliocene interval that implies that two distinct modes of sedimentation from particular sources occurred. Figure continues next page.

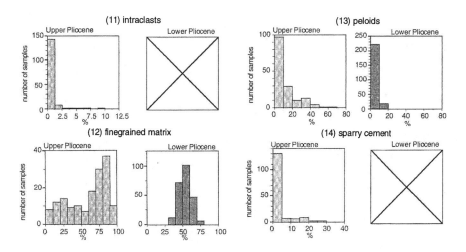

Fig. 14. contd.

foraminifers (-0.59) reflects the high amounts of peloids and foraminifers in the samples at the base of the coarse-grained intervals. The correlation matrix thus describes the compositional differences between coarse-grained and fine-grained sediments, and the transitions, that were observed in the Upper Pliocene interval.

Principal component analysis led to the calculation of unrotated factors (Table 3) that show relatively low eigenvalues (i.e. contributions to the total variability). This is characteristic for complex systems. The non-determinable grains have been included in statistical analyses as they possibly represent aragonitic skeletal grains such as coral fragments. In the selected Upper Pliocene interval, the first three factors explain 52.7% of the total variation. The first factor (28.0%) is dominantly controlled by sparry cement, cortoids, benthic foraminifers, and intraclasts (positive loadings), and by fine-grained matrix (negative loading). It characterizes coarse-grained, matrix-poor deposits derived from platform margin environments. The second factor (13.4%) is positively loaded with non-determinable components (that are thought to represent skeletal grains), peloids, and benthic foraminifers, and is negatively loaded with cortoids, and red algae. This factor represents the peloid-rich sediments at the top of the coarsening-up trends below the coarse-grained intervals. The third factor (11.3%) is positively loaded with *Halimeda* and bryozoans, and negatively loaded with sparry cement, cortoids, and peloids. This factor might represent the *Halimeda*-rich, fine-grained packstones at the top of the coarse-grained intervals.

Upper Pliocene	Mean	Std.Dev.	Std.Error	Minimum	Maximum	Variance	Coef.Var.	Skew
(1) Halimeda	0.78	2.03	0.16	0.0	15.5	4.10	2.61	4.827
(2) red algae	0.33	2.01	0.16	0.0	21.0	4.06	6.06	8.563
(3) molluscs	0.20	0.37	0.03	0.0	2.0	0.14	1.86	2.128
(4) bryozoans	0.09	0.51	0.04	0.0	6.0	0.26	5.89	10.124
(5) benth. forams	1.59	2.45	0.20	0.0	15.5	5.99	1.54	2.762
(6) plankt. forams	0.31	0.48	0.04	0.0	2.5	0.23	1.51	1.894
(7) ostracods	0.22	0.39	0.03	0.0	1.5	0.15	1.78	1.757
(8) echinoderms	0.15	0.33	0.03	0.0	1.5	0.11	2.21	2.479
(9) non-determinable	18.36	8.79	0.71	1.5	46.0	77.22	0.48	0.669
(10) cortoids	5.16	12.60	1.01	0.0	56.5	158.64	2.44	2.807
(11) intraclasts	0.28	1.12	0.09	0.0	9.0	1.24	3.97	5.565
(12) fine-gr. matrix	59.23	26.96	2.17	1.0	96.5	726.82	0.46	-0.603
(13) peloids	10.65	13.02	1.05	0.0	63.5	169.41	1.22	1.670
(14) sparry cement	2.65	5.68	0.46	0.0	26.0	32.28	2.15	2.320

Table 1. Summary statistics of the 14 point count-groups counted in the selected 155 samples from the Upper Pliocene interval. A total of 200 points were counted in each thin section.

Upper Pliocene	(1)	(2)	(3)	(4)	(5)	(6)	(7)	(8)	(9)	(10)	(11)	(12)	(13)	(14)
(1) Halimeda	1.00	0.10	0.36	0.49	0.13	0.16	0.02	0.31	0.07	0.16	0.00	-0.15	-0.13	-0.02
(2) red algae	0.10	1.00	0.08	0.15	0.06	0.37	0.06	0.15	-0.03	0.36	0.23	-0.28	-0.07	0.24
(3) molluscs	0.36	0.08	1.00	0.17	0.33	0.10	0.05	0.31	0.30	0.26	0.24	-0.41	0.10	0.25
(4) bryozoans	0.49	0.15	0.17	1.00	0.21	0.18	0.00	0.03	-0.04	0.37	0.06	-0.24	-0.07	0.12
(5) benth. forams	0.13	0.06	0.33	0.21	1.00	0.18	0.19	0.26	0.50	0.13	0.48	-0.59	0.31	0.35
(6) plankt. forams	0.16	0.37	0.10	0.18	0.18	1.00	0.03	0.12	0.10	0.10	0.01	-0.27	0.20	0.03
(7) ostracods	0.02	0.06	0.05	0.00	0.19	0.03	1.00	0.05	0.21	0.00	0.07	-0.10	-0.04	0.05
(8) echinoderms	0.31	0.15	0.31	0.03	0.26	0.12	0.05	1.00	0.15	0.27	0.12	-0.36	0.08	0.30
(9) non-determi.	0.07	-0.03	0.30	-0.04	0.50	0.10	0.21	0.15	1.00	-0.22	0.23	-0.39	0.23	-0.05
(10) cortoids	0.16	0.36	0.26	0.37	0.13	0.10	0.00	0.27	-0.22	1.00	0.29	-0.64	1.40	0.80
(11) intraclasts	0.00	0.23	0.24	0.06	0.48	0.01	0.07	0.12	0.23	0.29	1.00	-0.41	0.00	0.40
(12) fine-gr. matrix	-0.15	-0.28	-0.41	-0.24	-0.59	-0.27	-0.10	-0.36	-0.39	-0.64	-0.41	1.00	-0.61	-0.72
(13) peloids	-0.13	-0.07	0.10	-0.07	0.31	0.20	-0.04	0.08	0.23	1.40	0.00	-0.61	1.00	0.17
(14) sparry cement	-0.02	0.24	0.25	0.12	0.35	0.03	0.05	0.30	-0.05	0.80	0.40	-0.72	0.17	1.00

Table 2. Correlation matrix describing the inter-relation of the components present in the selected Upper Pliocene interval. Non-transformed point-count values of 155 samples are included. Correlations that exceed 0.50 are underlined.

Upper Pliocene	Factor 1	Factor 2	Factor 3
(1) Halimeda	0.32	-0.28	<u>0.76</u>
(2) red algae	0.39	-0.37	-0.02
(3) molluscs	0.56	0.09	0.34
(4) bryozoans	0.37	-0.41	0.48
(5) benth. forams	0.67	0.45	0.08
(6) plankt. forams	0.33	-0.02	0.26
(7) ostracods	0.15	0.24	0.12
(8) echinoderms	0.50	-0.03	0.19
(9) non-determinable	0.35	<u>0.73</u>	0.26
(10) cortoids	0.68	<u>-0.59</u>	-0.29
(11) intraclasts	0.55	0.10	-0.21
(12) fine-grained matrix	<u>-0.92</u>	-0.14	0.20
(13) peloids	0.35	0.50	-0.26
(14) sparry cement	<u>0.73</u>	-0.27	<u>-0.49</u>

Table 3. First three unrotated factors describing 52.7% of the variations in the composition of the selected Upper Pliocene interval. Highest positive and negative values for each factor are underlined.

4.2.3.2
Statistics of the Lower Pliocene Composition

In the selected succession in the Lower Pliocene, numerical analyses of the 238 samples revealed the following statistical behavior (four duplicate samples of the original 242 samples that have been point counted were excluded from statistical analyses). Of the frequency histograms, only few show an asymptotic shape (e.g. planktic foraminifers), others exhibit a bell-curve (e.g. fine-grained matrix; Fig. 14). These latter histograms are characterized by a low skewness (0.11, -0.42, resp.; Table 4). This indicates that these point-count groups have a similar frequency in most samples, and the "Gaussian tails" are statistical deviations of the normal distribution rather than indicative of systematic differences in the input. Components found to be most indicative in the selected Upper Pliocene succession (clasts, cortoids, sparry cement) are absent from the Lower Pliocene interval examined.

The coefficient of variation, typifying the spatial restriction of the facies belt a point-count group originates from, shows the highest values for red algae, bryozoans, and *Halimeda*. These groups are present in the selected Lower Pliocene interval in amounts too low to be statistically significant. The point-count groups present in sufficient amounts are characterized by low coefficients. They seem to be derived from facies belts that are spatially not very well defined as, for example, a broad platform margin that interfingers with a gentle foreslope or with a backward lagoon.

Lower Pliocene	Mean	Std.Dev.	Std.Error	Minimum	Maximum	Variance	Coef.Var.	Skew.
(1) Halimeda	0.01	0.06	0.00	0.0	0.5	0.00	8.87	8.74
(2) red algae	0.00	0.03	0.00	0.0	0.5	0.00	15.43	15.33
(3) molluscs	0.03	0.14	0.01	0.0	1.0	0.02	4.84	5.25
(4) bryozoans	0.01	0.07	0.00	0.0	1.0	0.01	11.48	12.30
(5) benth. forams	2.99	1.81	0.12	0.0	10.0	3.29	0.61	1.09
(6) plankt. forams	1.30	1.96	0.13	0.0	20.5	3.84	1.50	6.06
(7) ostracods	0.15	0.34	0.02	0.0	2.0	0.12	2.36	2.57
(8) echinoderms	0.08	0.24	0.02	0.0	2.0	0.06	3.17	4.30
(9) non-determinable	38.47	8.60	0.56	12.5	60.5	73.95	0.22	-0.40
(12) fine-gr. matrix	53.13	8.50	0.55	31.5	74.5	72.21	0.16	0.10
(13) peloids	3.85	3.12	0.20	0.0	14.5	9.75	0.81	1.29

Table 4. Summary statistics of the 11 point count-groups that compose the selected 238 samples from the Lower Pliocene interval. The three additional point-count groups that are shown for the Upper Pliocene interval, but are missing in this table (cortoids, intraclasts, sparry cement) are absent from the Lower Pliocene or were only found once. A total of 200 points were counted in each thin section.

Lower Pliocene	(1)	(2)	(3)	(4)	(5)	(6)	(7)	(8)	(9)	(12)	(13)
(1) Halimeda	1.00	-0.01	-0.02	-0.01	-0.01	-0.01	-0.05	-0.04	0.10	-0.11	0.02
(2) red algae	-0.01	1.00	-0.01	-0.01	0.48	0.12	-0.03	0.25	0.05	-0.11	0.07
(3) molluscs	-0.02	-0.01	1.00	-0.02	-0.02	-0.02	0.04	0.06	-0.09	0.11	-0.08
(4) bryozoans	-0.01	-0.01	-0.02	1.00	-0.06	-0.06	0.13	-0.03	-0.01	-0.02	0.09
(5) benth. forams	-0.01	0.48	-0.02	-0.06	1.00	0.23	-0.12	0.06	-0.33	0.03	0.11
(6) plankt. forams	-0.01	0.12	-0.02	-0.06	0.23	1.00	-0.08	0.63	-0.15	-0.14	0.00
(7) ostracods	-0.05	-0.03	0.04	0.13	-0.12	-0.08	1.00	0.03	0.10	-0.09	-0.03
(8) echinoderms	-0.04	0.25	0.06	-0.03	0.06	0.63	0.03	1.00	0.03	-0.21	-0.04
(9) non-determin.	0.10	0.05	-0.09	-0.01	-0.33	-0.15	0.10	0.03	1.00	-0.86	-0.14
(12) fine-gr. matrix	-0.11	-0.11	0.11	-0.02	0.03	-0.14	-0.09	-0.21	-0.86	1.00	-0.25
(13) peloids	0.02	0.07	-0.08	0.09	0.11	0.00	-0.03	-0.04	-0.14	-0.25	1.00

Table 5. Correlation matrix describing the inter-relation of the components present in the selected Lower Pliocene interval. Untransformed point-count values of 238 samples are included. Correlations that exceed 0.50 are underlined.

The coefficient of variation, typifying the spatial restriction of the facies belt a point-count group originates from, shows the highest values for red algae, bryozoans, and *Halimeda*. These groups are present in the selected Lower Pliocene interval in amounts too low to be statistically significant. The point-count groups present in sufficient amounts are characterized by low coefficients. They seem to be derived from facies belts that are spatially not very well defined as, for example, a broad platform margin that interfingers with a gentle foreslope or with a backward lagoon.

In the selected Lower Pliocene interval, correlations between the point-count groups generally are very weak (Table 5). The highest correlation observed is the high negative correlation between fine-grained matrix and the non-determinable grains (-0.86). As the fine-grained matrix and the non-determinable, diagenetically altered and/or small-sized components are by far the dominant constituents of the samples from the interval examined in the Lower Pliocene, their quantity is expected to correlate negatively. A second strong correlation is the positive correlation between planktic foraminifers and echinoderms (0.63), that reflects the joint occurrence of these two groups in the lowermost samples of the interval examined. All other groups show neither strong positive nor strong negative correlations and thus show a rather neutral behavior.

Of the total variation, 46.5% are explained by the first three factors (Table 6). The dominance of the fine-grained matrix and the non-determinable components is also reflected in the factor analysis. The first factor (18.5%) is loaded negatively with non-determinable components and positively with the fine-grained matrix. It represents the packstone-wackestone alternations that characterize the Lower Pliocene interval. The second factor (16.6%) is positively loaded with planktic foraminifers and echinoderms, and is negatively loaded with fine-grained matrix, and subordinately with ostracods and bryozoans. It is thought to characterize the varying influence of the platform top (various skeletal grains), and the open marine environment (planktic foraminifers, echinoderms). The third factor (11.4 %) is typified by positive loadings of peloids and benthic foraminifers, and negative loads of mollusks and non-determinable grains. It reflects the slight input cycles observed in peloids and benthic foraminifers, and thus possibly the productivity and/or the export from the platform top.

Lower Pliocene	Factor 1	Factor 2	Factor 3
(1) Halimeda	-0.18	-0.08	0.26
(2) red algae	-0.25	0.35	-0.04
(3) molluscs	0.16	0.04	-0.46
(4) bryozoans	-0.04	-0.14	0.05
(5) benth. forams	0.26	0.48	0.43
(6) plankt. forams	-0.16	0.85	-0.09
(7) ostracods	-0.17	-0.19	-0.36
(8) echinoderms	-0.35	0.76	-0.31
(9) non-determinable	-0.89	-0.35	-0.09
(12) fine-grained matrix	0.94	0.01	-0.21
(13) peloids	-0.10	0.10	0.71

Table 6. First three unrotated factors describing 46.5% of the variations in the composition of the selected Lower Pliocene interval. Highest positive and negative values for each factor are underlined.

4.2.4
Foraminiferal Assemblages

Foraminiferal assemblages exhibit the mixed shallow-water/open marine characteristics typical for periplatform sediments. As determined by point counting, benthic foraminifers account for an average of 2.7% in the selected Upper Pliocene interval (with two maxima below the bases of the two coarse-grained layers, a very prominent one of 23.5% at 257.81 mbmp and a subordinate one of 4.5% at 225.70 mbmp). In the selected Lower Pliocene succession, on the average 3.0 % of the sediment volume are composed of benthic foraminifers. A maximum of 10% is found at 502.46 mbmp, but small-scale variations exceed the large-scale trends. Planktic foraminifers are present with abundances of up to 2.5% in the Upper Pliocene succession (average 0.4%). The fraction of planktic foraminifers is higher in the Lower Pliocene interval examined (1.3% average), and at the base of the interval (509.47 mbmp), a distinct peak of 20.5% is observed.

Quantitative analysis of foraminifers by determining the abundance of specimens/cm^2 complemented the point count analysis. It revealed a more detailed image of the distributions of the various groups of foraminifers that have been considered: (1) Miliolidae and subordinately Soritidae, (2) *Rosalina*, (3) Amphisteginidae, (4) Textulariidae and *Reusella*, (5) other benthic foraminifers, mainly Rotaliidae, and finally (6) planktic foraminifers.

4.2.4.1
Distribution of Foraminifers in the Upper Pliocene

The foraminiferal assemblages (determined as specimens/cm$^{2)}$ exhibit different sources. Shallow-water and deeper-water benthic foraminifers occur together with planktic specimens (Fig. 15).

The benthic foraminifers exhibit specific vertical variations in the distribution. Miliolids are dominated by very small specimens (around 0.2 mm in diameter) with mostly rotund, thin tests that occur throughout the entire succession. While numbers of around 10 specimens/cm^2 are common, miliolids are most abundant in the uppermost part of the fine-grained deposits where they reach quantities exceeding 200 specimens/cm^2. Large tests of soritids, that similar to rotund miliolids are typical for very shallow water, are absent from most samples. Nevertheless they are infrequently observed in and below coarse-grained deposits. Large amphisteginid tests are usually absent from the fine-grained deposits. They are almost exclusively found in coarse-grained deposits and show maximum quantities of 20 specimens/cm^2 at the top of the coarse-grained intervals. *Rosalina* exceeds 100 specimens/cm^2 in some samples while frequently 5 to 10 specimens/cm^2 are found. Peaks in their distribution are found at the top of the fine-grained sediments and in the coarse-grained deposits. Textulariids and *Reussella*, that are typical for slope environments, are observed in low amounts in the coarse-grained intervals. Textulariids also sporadically occur in fine-grained samples. Similar to *Rosalina*, the group of other benthic foraminifers, dominated by rotaliids, shows low numbers in the fine-grained deposits. Towards the top of the fine-grained deposits this group increases in numbers and reach maxima around 50 specimens/cm^2.

Planktic foraminifers (mostly Globigerinidae, rarely Globorotaliidae) are constantly present in the fine-grained deposits of the selected Upper Pliocene section with two to five specimens/cm^2. In the uppermost part of the fine-grained interval and in the coarse-grained sediments they are most abundant, reaching up to 40 specimens/cm^2.

In summary, the trends in the selected interval in N22 are described as follows: In the upper part of the fine-grained interval, the abundance of benthic groups such as miliolids, *Rosalina*, and rotaliids increases considerably. At the same time, planktic foraminifers become more abundant. In the coarse-grained intervals, the aforementioned benthic foraminifers are present in somewhat lower numbers, while the planktic foraminifers show varying abundances. Planktic foraminifers are present in high numbers (up to 40 specimens per cm^2) in the coarse-grained interval around 256 mbmp, whereas they are rare in the coarse-grained interval around 221 mbmp. At the top of the coarse-grained intervals, amphisteginids become an important group and show a distinct peak. This peak is also observed in the slope group (textulariids and *Reussella*). The overlying lower fine-grained intervals (above 220 and 255 mbmp) again are characterized by low numbers of benthic and planktic foraminifers.

4.2.4.2
Distribution of Foraminifers in the Lower Pliocene

In the interval examined from the Lower Pliocene, the quantities of some benthic foraminifer groups are distinctly lower than in the selected succession from N22 (Fig. 16). Miliolids and soritids do not exceed 15 specimens/cm^2 and show no clear trend. Amphisteginids are extremely rare. *Rosalina* averages around 15 specimens/cm^2, and exhibits a maximum of 60 specimens/cm^2 in the lower part of the interval examined (504 mbmp). Other benthic foraminifers, mainly rotaliids, show a peak exceeding 150 specimens/cm^2, parallel to the maximum of *Rosalina*. Above this peak, varying abundances of this group between 10 to 120 specimens/cm^2 lack a clear trend. Benthic foraminifers typical for slope environments (*Reussella*, textulariids) repeatedly occur in low numbers throughout the succession.

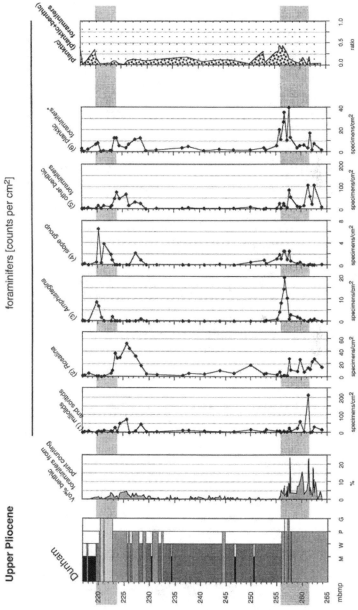

Fig. 15. Abundance of foraminifers in the selected Upper Pliocene interval (in specimens per cm²). Note the distinctly higher numbers in benthic foraminifers in the coarse-grained intervals. Left column shows classification of the samples after Dunham. Second column is the volumetric portion of benthic foraminifers derived from point counting, see figure 13. In the column with the slope-derived foraminifers (4), the dashed line marks the textulariids that contribute to the curve. Right column shows the relative amounts of planktic foraminifers per cm² (heavy stipples) versus benthic foraminifers per cm² (light stipples).

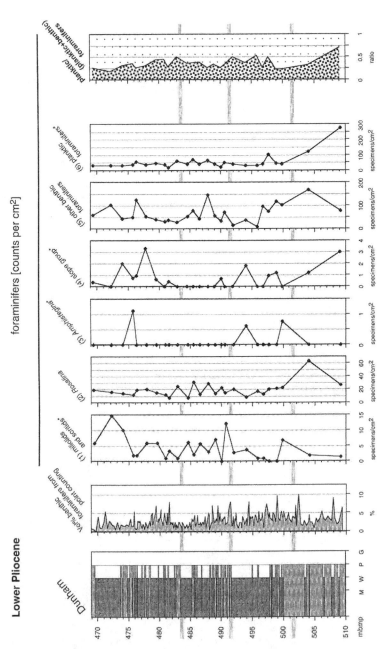

Fig. 16. Abundance of foraminifers in the selected Lower Pliocene interval (in specimens per cm²). Left column shows classification of the samples after Dunham. Second column is the volumetric portion of benthic foraminifers derived from point counting, see figure 13. Right column shows the relative amounts of planktic (heavily stippled) versus benthic foraminifers (light stipples). Columns with scales differing from those of the corresponding column in figure 15 are marked with an asterisk.

In the selected Lower Pliocene interval, planktic foraminifers are more abundant than in the selected Upper Pliocene interval, and average around 50 specimens/cm^2. At the bottom of the interval examined (below 509 mbmp), numbers of planktic foraminifers increase to a maximum of 275 specimens/cm^2.

Overall, the distribution patterns observed in the Lower Pliocene interval are less clear than in the Upper Pliocene interval. An obvious trend is the exceptionally high peak of planktic foraminifers at the base of the interval. The variations in the abundances of the different benthic groups, however, show no systematic correlations, and do not seem to correlate to the ambiguous cyclicities of peloids and total benthic foraminifers as observed in component analysis (point counting; Fig. 13 and second column in Fig. 16).

4.2.4.3
Environmental Significance of Foraminiferal Assemblages

Murray (1973) has invented a practical method to deduce environmental information from foraminiferal assemblages. The method is based on the fact that all modern foraminiferids belong to one of the three suborders Miliolina, Rotaliina, and Textulariina (Loeblich and Tappan, 1964). According to Murray (1973), environmental properties are expressed in the relative abundances of the three large suborders of benthic foraminifers. The relative abundances can be plotted in a triangular diagram (Fig. 17A). Numerous studies in various modern environments have shown that samples of specific environments cluster in defined fields in the triangle. For example hypersaline environments plot close to the corner of the triangle that represents 100% Miliolina, and extend down to more Rotaliina-dominated areas. Hyposaline environments plot along the axis that connects Rotaliina with Textulariina.

This method has been developed for *in situ* assemblages, therefore an application to allochthonous periplatform sediments has to be interpreted carefully. Nevertheless the separation of the Upper and Lower Pliocene assemblages, and the trends observed between fine-grained and coarse-grained deposits within the Upper Pliocene samples suggests that an original signature has been preserved.

The samples from the selected Pliocene intervals plot in different clusters (Fig. 17B). The Upper Pliocene samples are concentrated at the leg of the triangle that connects the Miliolina with the Rotaliina. Foraminiferal assemblages of the fine-grained deposits plot predominantly in the field that characterizes hypersaline lagoons. Coarse-grained deposits, in contrast, are found in the field of normal-marine lagoons. Lower Pliocene samples are concentrated near the corner that represents 100% Rotaliina. This area is less clearly defined, because here the fields of hypersaline lagoons, normal saline lagoons, and marine shelves overlap.

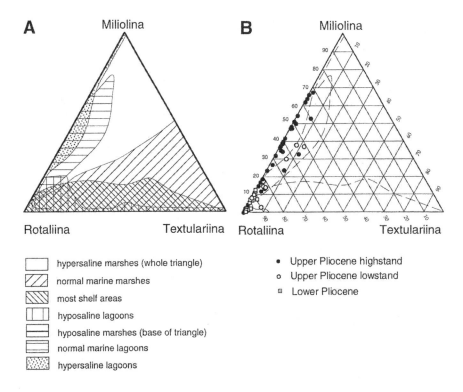

Fig. 17. (A) Environmental information drawn from *in situ* foraminiferal assemblages (Murray, 1973). The relative abundances of the three large groups of present-day benthic foraminifers (Rotaliina, Miliolina, Textulariina) reflect the environmental conditions with respect to salinity, restriction, and water depth. **(B)** Allochthonous foraminiferal assemblage in samples from CLINO plotted in the triangle of Murray (1973). Although these are no *in situ* assemblages, they yield information on the environmental conditions of the source area.

4.2.5
Microfacies Descriptions

Fourteen microfacies types (MF) are defined based on statistical results of the component analysis, on Dunham and Folk classifications (as noted in parenthesis), as well as on quantitative evaluation of foraminiferal assemblages, semiquantitative, and qualitative observations (bioturbation, lamination, porosity). They range from coarse grainstones with clearly platform-derived bioclasts to mudstones with low amounts of foraminifers. As the separation of most microfacies types is defined on relative abundances, there are no sharp boundaries between some of them. The microfacies found in the upper and the Lower Pliocene interval differ and show no overlap, so that the descriptions of the microfacies types present are split into two parts.

All the samples from the selected intervals have been deposited in the same facies zone 3 (shelf slope, clinothem) of Wilson (1975), therefore a classification following the standard microfacies types of Wilson (1975) was renounced.

4.2.5.1
Upper Pliocene Microfacies

Typical aspects of the Upper Pliocene section. The thick, fine-grained deposits are rich in sedimentary matrix and contain high amounts of fine-grained biodetritus. They are usually well to moderately sorted. Mudstones are typical for the basal unit of the fine-grained intervals (Fig. 12). Above the mudstones, peloid-bearing pack- to wackestones and skeletal wackestones alternate over most of the interval. At the top of the relatively uniform fine-grained interval, a transition to the coarse-grained intervals can occur that is composed of peloid-rich packstones and miliolid packstones. In this position, biodetritus-rich sediments also are found.

Above fine-grained deposits, more variable coarse-grained intervals follow. Cortoids are abundant in the coarse-grained intervals, but also green and red algae are common. Sorting of the coarse-grained sediments is generally poor, and broken bioclasts, such as the massive tests of amphisteginids, are frequently found.

MF 1: Cortoid Grainstones (n=14). (Grainstones; intrasparite, pelintrasparite, biointrasparite, some poorly washed)

MF 1 (Fig. 18A; Plate 5A) occurs in two intervals within the selected section of N22. Between 220.26 and 223.24 mbmp, eight samples belong to MF 1, while between 256.49 and 259.08 mbmp, six samples show MF 1 characteristics.

Cortoid grainstones contain up to 50% cortoids. Subordinate components are peloids and red algae. Benthic foraminifers are dominated by rotaliids and amphisteginids. This coarse-grained assemblage is thus strongly dominated by shoal, rim, and upper slope components. It is typified by the first of the three factors of the factor analysis, that is dominated by sparry cement and cortoids.

Sorting is moderate in these grainstones. Sand-sized components show maximum grain-sizes of 2.0 mm, but smaller grains (around 0.2 mm, medium-sized sand) dominate. Compaction is absent from the usually tightly cemented intrasparites. Bioturbation is not visible in these coarse-grained sediments.

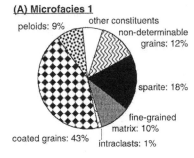
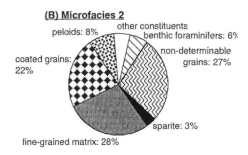

Fig. 18. Average composition of (A) microfacies 1 and (B) microfacies 2. Percentages from point-counting.

MF 2: Cortoid Packstones (n=4). (Packstones; packed intrabiomicrite, packed biointramicrite, packed pelintramicrite)

MF 2 (Fig. 18B; Plate 5B) occurs interlayered with MF 1 in the interval between 256.49 and 259.08 mbmp, but is absent from the upper coarse-grained interval that yields samples belonging to MF 1 (at about 222 mbmp).

This microfacies is typified by its fine-grained matrix (micrite and microspar), and large amounts of cortoids. Beside cortoids, benthic foraminifers (dominantly rotaliids), strongly altered non-determinable grains, and small amounts of other components occur.

Moldic porosity is low, and indications of compaction and bioturbation are absent. The composition is similar to MF 1, but grain-sizes in MF 2 are slightly smaller (0.2 mm, up to 1.3 mm). Sediments of MF 2 are poorly sorted. Rarely, single cobble-sized components of up to 8.0 mm in length (e.g. bryozoans) occur.

MF 3: Halimeda-rich Biomicrites (n=8). (Packstones to wackestones; biomicrite, packed pelbiomicrite)

Whereas MF 3 (Fig. 18C; Plate 5C) is rare in the coarse-grained interval around 220 mbmp, it is common in the upper part (256.03 to 256.82) of the lower MF 1-bearing interval.

MF 3 is composed of fine-grained matrix, cortoids, peloids, and grains of biotic origin. Unlike MF 2, it contains large amounts of *Halimeda* plates (11%). MF 3 is characterized by poor sorting. Because of their shape, large *Halimeda* plates (coarse sand; up to 10 mm in length) are frequently oriented parallel to the bedding. Large Amphisteginids are common, while the bryozoan *Adeona* is infrequently found. Non-determinable, strongly diagenetically altered grains are common in these poorly sorted calcarenites. The third of the three factors describing the variation of the Upper Pliocene samples typifies the composition of MF 3.

Bioturbation is indicated in thin section by spots of different grain-sizes. Porosity is highly variable. In some samples *Halimeda* plates are leached and contribute to moldic porosity. In other samples, some interparticle porosity is preserved. Indications for compaction are absent.

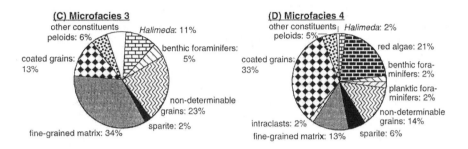

Fig. 18 contd. (C): average composition of microfacies 3. (D): average composition of microfacies 4. Percentages from point-counting.

MF 4: Nodule-rich Biomicrites (n=1). (Packstones, packed biomicrite)

MF 4 typifies one sample (256.79 mbmp; Fig. 18D; Plate 5D) in the selected Upper Pliocene interval that is located in the upper part of the lower MF 1-bearing interval. Nevertheless, the extraordinarily high amounts of red algae nodules justifies recognition of a separate microfacies type.

MF 4 belongs to the cortoid-bearing microfacies. It is similar to MF 2 but is distinguished by high amounts of red algal nodules (21%). Sorting is poor, and large red algal nodules (coarse sand-sized; up to 12 mm in length), large Amphisteginids and other larger bioclasts, are found together with finer-grained biodetritus. Some amphisteginid tests are preserved as fragments.

MF 5: Miliolid Packstones (n=12). (Packstones, subordinately wackestones; packed pelbiomicrite, also packed intrabiomicrite, biomicrite, poorly-washed pelbiosparite)

Samples exhibiting characteristics of MF 5 (Fig. 18E; Plate 5E) are found exclusively at the base (257.71 to 263.80 mbmp) of the lower coarse-grained interval.

These bioclast-dominated packstones are conspicuous due to their high amounts of miliolids. Up to 210 specimens/cm^2 were counted (261.39 mbmp). Almost exclusively, small rotund forms are found (around 0.1 mm in diameter). Besides miliolids, tests of *Rosalina* are observed. The packstones are fine-grained and moderately to well sorted, with most constituents having a size similar to or smaller than the miliolid tests. Moldic porosity is low to absent. Some samples are bioturbated as is indicated by locally deviating color and grain sizes in thin section. In the frequency histogram, showing the distribution of the percentages of (12) fine-grained matrix, this microfacies would plot together with MF 1, 2, 3, and 4 on the left peak, which depicts the matrix-poor samples (Fig. 14).

MF 6: Mixed Pack- to Wackestones (n=55). (Packstones. wackestones; (packed) biopelmicrite, (packed) pelbiomicrite, subordinately biogene-rich Pelsparite)

This fine-grained microfacies is common in the Upper Pliocene succession. It preferentially occurs between 223.29 and 250.15 mbmp, and between 261.06 mbmp and the base of the investigated Upper Pliocene interval. This corresponds to a position below the coarse-grained intervals that are defined by the occurrence of microfacies MF 1 to 4. Sediments of MF 6 are less common in the interval between 234.60 and 244.00 mbmp within the highstand deposits, but are present throughout the entire Upper Pliocene interval examined.

The fine-grained matrix (micrite and microspar) of the samples of MF 6 (Fig. 18F; Plate 5F) varies around 50% of the total volume. Peloids are abundant and make up about half of the components that show grain sizes of around 0.6 mm in diameter. A large portion of the remaining components consists predominantly of strongly altered, fine-grained biodetritus (0.1 mm in diameter). Benthic foraminifers occur in varying abundances and assemblages. Some Upper Pliocene samples contain up to 102 specimens of rotaliids/cm^2 and up to 52 specimens of *Rosalina*/cm^2. Miliolids and planktic foraminifers also are present. The grains identified point to a strong influence from the platform interior.

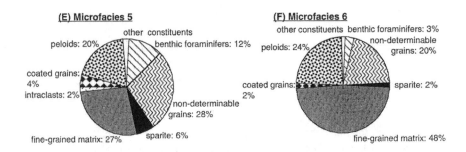

Fig. 18 contd. (E): average composition of microfacies 5. (F): average composition of microfacies 6. Percentages from point-counting.

Bioturbation is frequently observed. Moldic porosity is moderate to low. Indications of compaction vary. While they are absent from some samples, other samples appear laminated as a result of strong compaction. In these samples, peloids are strongly deformed.

MF 7: Peloid Packstones (n=3). (Packstones; packed pelmicrite, biogene-rich)

Peloid packstones of MF 7 (Fig. 18G; Plate 5G) are rare and occur exclusively at the base of the lower coarse-grained interval (261.70-261.90 mbmp) in the selected Upper Pliocene succession.

MF 7 comprises samples in which the amount of peloids exceeds 49% of the constituents present. Besides the peloids and some fine-grained biodetritus, the most conspicuous components are foraminifers, dominated by rotaliids.

MF 7 is usually characterized by compactional features. Peloids are strongly deformed, sometimes leading to homogenization. Exclusively in MF 7, some broken foraminifer tests were observed. The samples appear laminated due to compaction, with fine biodetrital grains aligned to the lamination.

Microfacies 7 is reflected by the peloid-dominated second of the three first factors describing the variation found in the Upper Pliocene samples.

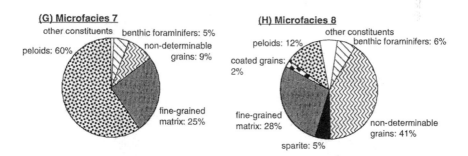

Fig. 18 contd. (G): average composition of microfacies 7. (H): average composition of microfacies 8. Percentages from point-counting.

MF 8: Biodetrital Packstones (n=4). (Packstone; packed biomicrite, peloid-rich, subordinate poorly washed peloid-rich biosparite and pelbiosparite)

MF 8 (Fig. 18H; Plate 5H) is rare in the Upper Pliocene succession, where it occurs below the lower coarse-grained interval and interbedded in these coarse-grained layers.

This microfacies is characterized by fine-grained, strongly altered biodetritus and lower amounts of peloids. The peloids partly occur in clusters of several tens of peloids. The peloids appear undeformed to slightly deformed. The sediment is moderately sorted. Rarely, poorly-washed, micrite-bearing sparites occur, but usually MF 8 exhibits a micritic matrix. Grain-sizes of the biodetritus average around 0.2 mm (fine-sand). Small specimens of foraminifers are common such as miliolids, rotaliids and globigerinids. Larger grains are also present, mainly amphisteginids and echinoderm fragments. As seen under the light microscope, moldic porosity is common in MF 8 and by far the dominant type of porosity.

MF 9: Biodetrital Wackestones (n=66). (Wackestone; biomicrite)

MF 9 (Fig. 18I; Plate 6A) occurs in the fine-grained intervals of the Upper Pliocene, mainly between the two coarse-grained intervals that are defined by the occurrence of MF 1.

This matrix-rich wackestone (on the average 75% fine-grained matrix; i.e. micrite and microspar) contains predominantly very fine sand-sized, diagenetically altered biodetritus (0.1-0.2 mm). Larger bioclasts and peloids are less common. In some samples, amphisteginids and *Halimeda* detritus attests an origin from the photic zone. Also, the strong diagenetic alteration indicates that the biodetritus was initially of metastable mineralogy, and thus most probably is derived from the platform top, margin, and upper slope. In many samples, the biodetritus is recrystallized, in others, moderate to high moldic porosity is present with molds around 0.1 to 0.2 mm in length. Sorting is good to moderate. Several samples appear laminated due to compaction.

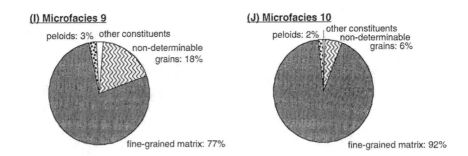

Fig. 18 contd. (I): average composition of microfacies 9. (J): average composition of microfacies 10. Percentages from point-counting.

MF 10: Mudstones (n=12). (Mudstones; micrite)

Sediments of MF 10 (Fig. 18J; Plate 6B) are found mainly on top of the coarse-grained intervals in the selected Upper Pliocene succession. The presence of

MF 10, especially above the upper coarse-grained interval, is clearly depicted by the sudden increase of fine-grained matrix in the point-counting curve (Fig. 12).

This component-poor sediment, consisting of more than 90% fine-grained matrix, contains some small rotaliids and *Rosalina* (around 0.2 mm in diameter), and fine biodetritus, and some planktic foraminifers. It appears homogeneous under the light-microscope. Bioturbation in some samples is indicated by slight color changes. Fine lamination and the orientation of fine biodetrital grains in other samples points to compaction. Porosity is low and mainly represented by primary interparticle porosity.

4.2.5.2.
Lower Pliocene Microfacies

Typical aspects of the Lower Pliocene section. The Lower Pliocene succession shows an irregular alternation of very fine-grained and well to moderately sorted biodetritus-rich wackestones, biodetritus-rich packstones of similar composition, and rare peloid-bearing packstones (Fig. 13). No obvious cyclicity is observed, although a general trend from matrix-poor towards matrix-rich sediments is present. The base of the interval examined is characterized by coarser-grained globigerinid packstones that are extremely rich in planktic foraminifers and contain abundant echinoderms. Sorting of this packstone is comparatively poor.

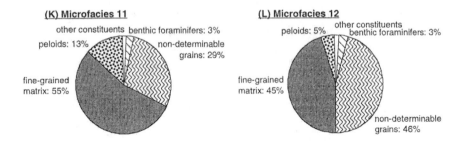

Fig. 18 contd. (K): average composition of microfacies 11. (L): average composition of microfacies 12. Percentages from point-counting.

MF 11: Bioclast-Peloid Packstones (n=5). (Packstones and wackestones. (packed) pelbiomicrite)

MF 11 (Fig. 18K; Plate 6C) occurs rarely in the middle to lower part of the Lower Pliocene succession. This microfacies, similar to MF 6, contains about 50% of fine-grained matrix (micrite and microspar), peloids, and non-determinable biodetritus. MF 11 typically contains higher amounts of non-determinable biodetritus (average of 34%). Grain sizes of the diagenetically altered biodetritus vary around 0.1 mm (very fine sand to silt). Peloids and planktic foraminifers are larger with grain sizes of 0.5 mm and 0.3 mm, respectively. Peloids frequently occur in clusters. Planktic foraminifers are abundant (around 50 specimens/cm^2), benthic foraminifers, mainly Rotaliids, occur in varying amounts.

MF 11 shows varying amounts of moldic porosity that originated from the dissolution of mineralogically metastable biodetritus. Compaction also varies, some samples show strongly deformed peloids, whereas in other samples the peloids appear undeformed.

MF 12: Bioclast Packstones (n=83). (Packstone; packed biomicrite, some peloid-rich)

MF 12 (Fig. 18L; Plate 6D) is ubiquitous in the lower part (498.81 to 508.56 mbmp) of the Lower Pliocene succession examined, but occurs throughout this interval.

Fine-grained, strongly altered biodetritus with uniform grain-sizes around 0.1 mm (very fine sand to silt) makes up almost half of these samples. Some larger peloids occur (around 0.5 mm) that are frequently arranged in clusters. Larger planktic foraminifers (up to 1.0 mm in diameter) are embedded in the fine-grained material. Benthic foraminifers occur in low to moderate amounts. MF 12 is characterized by fine grain sizes and moderate to good sorting.

Porosity varies strongly; some samples have low porosities, while in others most biodetritus is dissolved, leaving molds that make up to 43% of some samples (on the average 27%). Compaction also varies, as is indicated by variations in the deformation of peloids (Plate 6D), and by apparent lamination in some samples. (For a detailed description of these diagenetic features, the reader is referred to Chapter 5).

MF 13: Bioclast Wackestones (n=152). (Wackestone; biomicrite)

MF 13 (Fig. 18M; Plate 6E) is the most characteristic microfacies for the Lower Pliocene succession examined. More than half of the samples examined from this interval belong to MF 13. The fine-grained appearance and intense diagenetic alteration of the components are responsible for the monotonous appearance of large parts of the selected interval of N 19.

MF 13 resembles MF 12 in the constituents present. The amounts of fine-grained matrix, however, is higher (micrite and microspar amount to an average of 58%). Most of the very fine sand to silt-sized grains in these wackestones consist of strongly altered biodetritus (grain-sizes around 0.1 mm). Some peloids occur.

Foraminifers are dominated by planktic forms. Rotaliids occur in moderate amounts while miliolids are rare.

Moldic porosity and compaction, both being diagenetic features, vary (see Chapter 5). Moldic porosity averages around 21% and reaches up to 40% in some samples. Varying compaction of the wackestones (similar to the packstones of MF 12) has apparently no relationship to the microfacies type.

MF 14: Globigerinid Packstone (n=2). (Packstone; packed Biomicrite)

MF 14 (Fig. 18N; Plate 6F) occurs exclusively at the base of the selected Lower Pliocene interval (509.17 and 509.47 mbmp). This microfacies type is characterized by a strongly open marine signature with large amounts of planktic foraminifers (15.5-20.5% of the total area in thin section; up to 277 specimens/cm^2). Other common components are large benthic foraminifers (mainly rotaliids) and echinoderm fragments. MF 14 is typified by the second factor describing the Lower Pliocene compositional variability, which is positively loaded with planktic foraminifers and echinoderms.

Sorting is poor, e.g., fine-grained biodetritus is found among the planktic foraminifers that reach diameters exceeding 1.0 mm. Color changes of the matrix indicate slight bioturbation. Porosity is generally low, and moldic porosity does not exceed 5%.

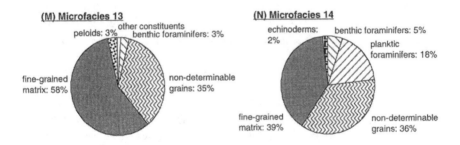

Fig. 18 contd. (M): average composition of microfacies 13. (N): average composition of microfacies 14. Percentages from point-counting.

4.3

Discussion - Compositional Signatures in Periplatform Sediments

4.3.1
Sediment Composition, Palynomorphs, and Sea-Level Fluctuations

The Pliocene periplatform carbonates from core CLINO that have been examined show a variety of compositional signatures. In most samples, large amounts of bank-derived constituents could be distinguished, pointing to an allochthonous nature of these slope sediments. Typical for periplatform deposits, the shallow-water derived constituents occur together with planktic foraminifers. Although the majority of the Upper and Lower Pliocene sediments are characterized by small-sized constituents and a fine-grained matrix, two thin, coarser-grained intervals in the selected Upper Pliocene interval are conspicuous. The concept of highstand shedding (Droxler et al., 1983; Droxler and Schlager, 1985; Reijmer et al., 1988, Haak and Schlager, 1989) implies that the high productivity of the shallow-water carbonate factory during sea-level highstands, as well as the cessation of banktop productivity during sea-level lowstands, potentially is reflected in the slope sediments. Thus, the variations in composition and in the abundance of bank-derived constituents in the successions examined could be a record of the changes in the environmental conditions of the source areas, namely on the platform top.

On the basis of lithologic examinations and seismic geometries, Kenter et al. (in press) and Eberli et al. (1997) interpreted the three major coarse-grained intervals from core CLINO (at around 540, 370, and 200 mbmp) as lowstand deposits of larger-scale sea-level cycles. For the smaller-scale variations examined in the present study, an analogous interpretation appears reasonable. The associations of palynomorphs offer direct proof for the sea-level influence on the fine-grained *versus* coarse-grained sediment composition in the Upper Pliocene succession. The fine-grained intervals are characterized by abundant normal marine to elevated saline dinoflagellate cyst associations (Appendix 2). These cysts point to conditions on the platform top similar to the present-day situation. In the coarse-grained intervals, in contrast, the palynomorph associations are typified by high amounts of pollen from *Pinus* and various angiosperms (Fig. 19). These pollen, once dropped in the water, behave like other silt-sized constituents. Assuming that the circulation patterns during the Pliocene were similar to the recent situation, then sediment transport was dominantly from the east to the west. Westward transport during the Pliocene is also indicated by seismic geometries that reveal thicker prograding clinoforms to the west than to the east of Great Bahama Bank (Fig. 3). This indicates that the dinoflagellate cysts as well as the pine pollen found in the periplatform sediments have probably been transported from the platform top of Great Bahama Bank to the western, leeward slope. The production of pine pollen requires subaerial exposure of at least parts of the platform top. Therefore it is concluded that during deposition of the coarse-grained sediments, large parts of Great Bahama Bank were subaerially exposed. Thereby the examination of palynomorphs proves to yield important information for the sedimentological interpretation of the slope sediments. The compositional patterns in the selected Upper Pliocene interval, thus confirmed to originate from sea-level fluctuations, once more verify the notion of "highstand shedding" (Droxler and Schlager, 1985).

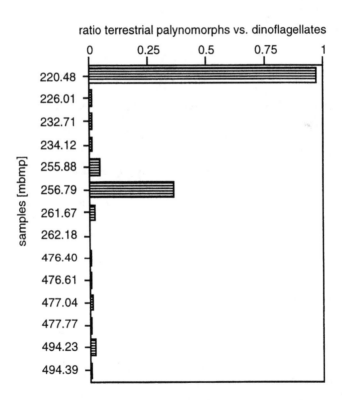

Fig. 19. Ratio of terrestrial pollen to dinoflagellate cysts. In the diagram, eight Upper and six Lower Pliocene samples are shown. Note the strikingly higher abundance of pollen in the two coarse-grained Upper Pliocene samples at 220.49 and 256.79 mbmp.

In the selected Lower Pliocene interval, sediments are fine-grained and micritic, and coarse-grained shallow-water derived deposits like those observed in the Upper Pliocene interval, are absent. Vague compositional cyclicities are only found in benthic foraminifers and peloids that exhibit three faint, stacked cycles. The palynomorph association of the selected Lower Pliocene interval appears rather uniform. Terrestrial palynomorphs are rare in the samples examined, therefore, an interpretation of environmental changes based on palynomorphs is not possible. Regarding the composition of the sediment, two samples at the base of the selected interval deviate from the general picture. They consist of comparably coarse-grained sediments. These sediments, however, are not platform-derived, but are of clearly open marine origin. The striking differences in the depositional record between the intervals selected in the Upper and Lower Pliocene have to be looked at closer.

4.3.2
Sedimentation of the Upper Pliocene Succession

The selected interval of Upper Pliocene sediments examined represents upper slope sediments deposited in a water depth of roughly 200 m. The distance to the margin, as marked by reefal deposits penetrated by core UNDA, was about 8.5 km (Eberli et al., 1997; in press).

The differentiated vertical distribution of constituents observed in the selected Upper Pliocene succession is interpreted to record sea-level changes at the time of deposition. The thin, matrix-poor, coarse-grained and poorly-sorted intervals are thought to represent sea-level lowstand deposits, whereas the thick, matrix-rich, fine-grained and well-sorted intervals are interpreted as highstand deposits.

The two coarse-grained intervals in the selected succession (around 220.0 to 224.0 mbmp and 256.0 to 261.5 mbmp; characterized mainly by MF 1 to MF 4) are characterized by poorly-sorted pack- to grainstones composed largely of rim to upper slope skeletal and non-skeletal grains. Among those are predominantly cortoids (especially in MF 1 and MF 2), *Halimeda* fragments (especially in MF 3), smaller amounts of benthic foraminifera, red algae, echinoderms and mollusks. On the present-day Bahamas, photic organisms such as *Halimeda* and specific benthic foraminifers like amphisteginids are known to prosper in rim to upper slope regions (Rose and Lidz, 1977; Ginsburg et al., 1991). These organisms are able to continue production and maintain export of calcareous material at least during minor sea-level lowstands by slightly shifting the facies belt downslope. Cortoids are almost exclusively present in the coarse-grained deposits. They are interpreted as allochthonous components sourced from the margin and upper slope of the carbonate platform. Assuming the majority of the micrite envelopes being of algal origin, the large numbers of cortoids probably originated from high-energy shoals (Gunatilaka, 1976; Flügel, 1982). As algal activity is also well known from more protected environments (Flügel, 1982), a contribution from lower energy lagoons of a more articulated shoreline is possible. The nodule-rich biomicrites (MF 4) are similarly interpreted to be shed from the rim to upper slope environment, where high water energy prevailed. Broken massive amphisteginid tests also point to resedimentation. The large amounts of photic organisms and the poor sorting (including very coarse to very fine sand-sized grains and mud-sized matrix constituents) indicate that the material is shed from the platform rim or upper slope probably as episodic mass flows during lowstand conditions.

Platform mud (predominantly aragonite needles) is rare to absent in the lowstand sediments, reflecting that sediment production within the platform interior was sharply reduced by subaerial exposure of the platform top. Selective bypassing of carbonate mud is considered inappropriate as an explanation for the virtual absence of fine-grained matrix in many of these sediments. During such a situation, density cascading (Wilson and Roberts, 1992, 1995) would still transport fine-grained material to the slope environment. Winnowing of the fine-grained material by currents, leaving a coarse-grained lag deposit also seems unlikely, as then a better sorting and breakage of thin-shelled planktic organisms would be expected. This, however, is not observed. Also, palynomorphs with sizes smaller than 30 µm, that easily could be removed by winnowing, are still present.

The association of palynomorphs that contains a distinctly higher fraction of pine pollen in the coarse-grained deposits, requires an emerged area that has been colonized by land vegetation. Numerous exposure surfaces are described from the Upper Pliocene platform top succession (Beach, 1982). Nevertheless, lithoclasts that would indicate erosion of an emerged platform top, are not observed in these lowstand deposits. The occurrence of lithoclasts could be restricted to turbidite gullies that were not penetrated, so that such gully infills were simply not recovered. Dravis (1996), though, has shown that emerged platforms potentially are lithified extremely rapidly by meteoric diagenesis (in the order of tens of years; "freezing"), thus inhibiting considerable erosion. Similarly, Schlager et al. (1994) have shown that early marine and meteoric diagenesis lead to a rapid lithification of the platform top that hinders mechanical erosion.

The locally higher, but strongly variable amounts of planktic foraminifers in the lowstand deposits of the Upper Pliocene interval could reflect a lowered sedimentation rate during sea-level lowstands, when the voluminous export from the platform interior ceases and episodic mass flows might have dominated.

A hardground at 256.20 mbmp mentioned by Melim et al. (in press-a) and Kenter et al. (in press) that would be located at the top of a coarse-grained interval examined in the present study, could not unequivocally be confirmed. This hardground is described to exhibit the micritic cementation typical for hardgrounds but to lack the actual bored surface that is interpreted to be eroded. In the present study, no definite proof to this hardground was found (for a discussion of the micritic cements see Chapter 5).

The fine-grained intervals in the selected Upper Pliocene succession are characterized by bioclastic to peloidal packstones, wackestones, and mudstones, attributed to the MF 6, MF 8, MF 9, and MF 10. These sediments are thought to represent sea-level highstand conditions. They are rich in material exported from the platform interior, namely aragonite needles. The extensive shallow-water carbonate factory was submerged at the time of deposition and thus able to produce and export great amounts of calcareous material. As Neumann and Land (1975) have shown, on the present-day Great Bahama Bank, vast volumes of excess material are exported from the banktop. During the Upper Pliocene, a similarly voluminous fine-grained platform input diluted the skeletal material shed from the rim and upper slope. High amounts of (diagenetically altered) originally metastable bioclastic material are observed in MF 6, MF 8, and MF 9. The varying amounts of peloids, that show a minor increase in abundance at 244 mbmp, could indicate subordinate changes in the conditions on the platform top or in the transport processes. The relatively good sorting and the dominantly small grain sizes indicate that the material settled from suspension. The absence of clear graded bedding, elsewhere frequently associated with settling from suspension, is probably a result of bioturbation that is observed in thin sections.

The micritic sediments of MF 3 at the top of the coarse-grained intervals are typified by the third factor of factor analysis, that is characterized by *Halimeda* fragments and bryozoans, and by the absence of sparite, cortoids, and peloids. These sediments could be the result of the early flooding event when most organisms were temporarily drowned. *Halimeda* seems to have given up shortly after, when mudstones have been deposited on top of MF 3. Mudstones (MF 10) are preferentially found directly above the coarse-grained intervals. The cause for this pulse of exported fine-grained material can only be speculated on. In earlier studies, aragonite percentages have been documented to show similar sudden, saw tooth-shaped, increases in younger intervals of Bahamian toe-of-slope sediments, although in the aragonite record, dissolution cycles could have modified the image (e.g. Kier and Pilkey, 1971; Droxler et al. 1983; Boardman et al., 1986; Reijmer et al., 1988). An unequivocal explanation for this pattern, however, has not yet been proposed. Below the "wall" of present day Great Bahama Bank, Wilber et al. (1990) observed Holocene subsurface sediments that were composed of 96% aragonite mud, where 70% of the bulk sediment is of clay size. These fine-grained, muddy deposits are thought to represent the Holocene flooding event. Possibly the micritic pulse observed in the selected Upper Pliocene interval from CLINO also reflects the flooding event that led to geologically instantaneous production of fine-grained material (chemical precipitation of aragonite needles and/or algal skeletal needles) whereas other organisms lagged behind. For the Holocene flooding event, lag-times between submergence and colonization by reef organisms ranged from 500 to 2000 years on different Indopacific islands (Montaggioni, 1988). Also, higher tidal energy in the deeper water of early flooding could have led to an enforced export of fine-grained material.

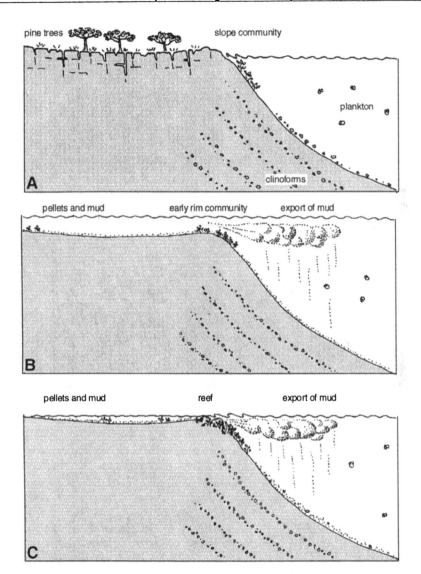

Fig. 20. Schematic sketch of the influence of sea-level fluctuations on the flat-topped platform in the Upper Pliocene. (A) during sea-level lowstands, the platform top is exposed and colonized by terrestrial plants. Sedimentation on the slope is dominated by slope biota. (B) Flooding of the platform leads to lag time deposition of muddy sediment. (C) During sea-level highstands, mud and pellets are exported from the platform interior to the slopes where they mix with margin-derived biota and planktic foraminifers.

In the selected Upper Pliocene interval from CLINO, the highstand deposits coarsen upcore, and packstones become more abundant. Late highstand deposits above 233 mbmp show a gradual increase in the amount of peloids. At the base of the interval examined (below 262 mbmp), also high amounts of peloids are observed that culminate in peloid packstones (MF 7). These peloid packstones precede the base of the overlying lowstand as marked by the appearance of cortoids. Peloids are a dominant constituent in surface sediments of present-day Bahamian platforms (Purdy, 1963). The increase of peloids associated with a coarsening-upwards trend might reflect the progressive decrease in the space for accommodation which is typical for late highstands when platforms are forced to build outwards. It could be associated with an increased export of coarser-grained material from the platform interior such as fecal pellets. Parallel, an increase in the export of some groups of benthic foraminifers occurs. Small, rotund miliolids occur in large numbers in the late highstand deposits. These miliolids are evidence for elevated saline conditions in the platform interior, coupled with the late sea-level highstand conditions. Miliolid packstones (MF 5) are restricted to the upper part of the fine-grained interval (260 mbmp) below the base of the coarse-grained interval.

The scenario deduced from microfacies distribution is supported by the foraminiferal assemblages. As seen in figure 17, the highstand sediments plot predominantly in the field of hypersaline lagoons, whereas the lowstand samples are found preferentially in the field of normal marine lagoons. During highstand conditions, large areas of the platform top where covered by shallow water, that could have had elevated salinities. During lowstands, the shallow-water areas were not available, so that mostly more open-marine organisms were shed downslope from the marginal areas of the platform.

Overall, the Upper Pliocene microfacies succession draws a picture of a typical flat-topped, steep-sided carbonate platform subjected to successive flooding, sea-level highstand, and subaerial exposure (Fig. 20). The sharp changes in composition observed in the Upper Pliocene succession correspond well with variations in sediment export as can be expected of a flat-topped, steep-sided carbonate platform. Flat-topped platforms are extremely sensitive to sea-level changes, as fluctuations of a few meters are sufficient to expose and flood the platform top (for an overview see Schlager et al., 1994). On the present-day Bahamas, the area available for carbonate production during highstands is two orders of magnitude larger than the area that is productive during lowstands (Schlager et al., 1994). A sea-level drop of 10 m is sufficient to expose most of the present day Great Bahama Bank, an area that corresponds to five times the size of the state of Schleswig-Holstein/Germany, or to the size of Scotland that covers about 78,000 km^2 (Fig. 21; Traverse and Ginsburg, 1966; Burchette and Wright, 1992). While the extensive shallow-water factory produced vast amounts of fine-grained material (mainly aragonite needles) during sea-level highstands that were exported to the slope, during sea-level lowstands much of the platform top was subaerially exposed, and productivity of aragonite needles and peloids ceased on the banktop. Neumann and Land (1975) have shown that on the recent Bahamas, productivity of the muddy facies in the platform interior is volumetrically by far superior to the reef and sand barriers. The produced excess of fine material is swept from the banks in suspension by waves and currents (Neumann and Land, 1975), and possibly density cascading took place (Wilson and Roberts, 1992; 1995). By these mechanisms, the fine-grained material settles from the water column and does not flow as a bottom current. The numerical analyses of the point-count results for the Upper Pliocene succession support the idea that two different modes of sediment production and transport prevailed during sea-level highstands and lowstands, respectively. This is best illustrated by the bimodal distribution of the percentages of fine-grained matrix (Fig. 15) where the peak at lower values represents the matrix-poor lowstand samples, while the peak at higher values corresponds to the matrix-rich highstand deposits.

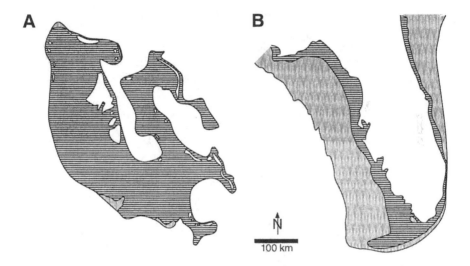

Fig. 21. Flat-topped carbonate platforms like Great Bahama Bank (A) are much stronger influenced by minor sea-level fluctuations than carbonate ramps like west of Florida (B). Whereas a 10 m drop of sea-level would expose most of the shallow-water carbonate factory of Great Bahama Bank, the same drop in sea-level would lead to emergence of a comparably narrow fringe west of Florida. On the gently dipping slope west of Florida, facies belts most probably would shift downslope with falling sea-level, and carbonate production would continue. Hatched areas: water depth < 10 m, stippled areas: water depth 10-100 m (simplified from Burchette and Wright, 1992, reprinted with the permission from Elsevier Science).

During sea-level lowstands, shallow-water production was relocated at the former margins, where on narrow shoals a less voluminous production of photic organisms (e.g. *Halimeda*, amphisteginids) and cortoids took place. Average sedimentation rates dropped, and more episodic bedload transport prevailed. As described from Belize (James and Ginsburg, 1979) and the Bahamas (Eberli and Ginsburg, 1989), a major sea-level fall in the order of 100 m causes flat-topped platforms to continue to grow as narrow ridges rimming the steep slope. The volume and composition of carbonate production on a carbonate platform can thus be changed abruptly by sea-level fluctuations.

The facies patterns observed in the Upper Pliocene periplatform sediments thus represent a straight-forward picture of the principle of highstand shedding *sensu* Droxler and Schlager (1985).

4.3.3
Sedimentation of the Lower Pliocene Succession

The selected Lower Pliocene interval is composed of lower slope sediments deposited at a depth of the order of 300 to 400 m. The distance to the shallow-water carbonate factory is estimated to 16 km (Eberli et al., 1997; in press).

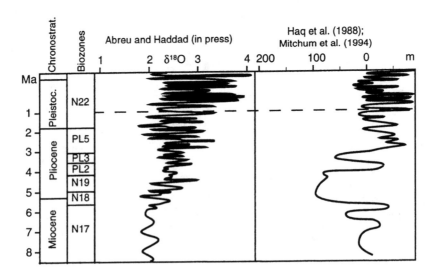

Fig. 22. Sea-level curve of Haq et al. (1988) and Mitchum et al. (1994), and stable oxygen isotope curve of Abreu and Haddad (in press). The latter is interpreted by these authors to be a direct image of sea-level fluctuations, that would have been much more numerous than previously thought. (From Abreu and Haddad, in press.)

The Lower Pliocene interval has been selected in spite of the rather uniform macroscopic appearance, because the gamma-ray log seemed to indicate an underlying higher-order cyclicity. Therefore, a possible sea-level signal has been presumed to be recorded in these sediments.

Sea-level fluctuations, however, are known from the entire Pliocene. Based on the correlation of isotope events and sequence stratigraphic events, Abreu and Haddad (in press) and Abreu and Anderson (in press) have argued that in the early Pliocene, high-frequency sea-level fluctuations (including 100 ka cyclicities) already occurred (Fig. 22). Also, sedimentary cycles in the platform-ward core UNDA (Kenter et al., in press), and discontinuity surfaces on the platform top at that time (Beach and Ginsburg, 1980) indicate sea-level fluctuations throughout the Pliocene. Based on sedimentation rates (Lidz and McNeill, 1995-a) the Upper Pliocene interval examined here covers a time span of about 100 ka, whereas the selected Lower Pliocene interval is thought to represent about 300 ka. Thus both successions are thought to have been subjected to environmental changes resulting from higher-frequency sea-level fluctuations. Nevertheless, the signature observed in the sedimentary record of both intervals differs considerably.

The sediment input features of the Lower Pliocene interval are remarkably different from the clear patterns observed in the selected Upper Pliocene interval. In the periplatform sediments of the selected Lower Pliocene interval, the drastic variations are absent, and more subtle changes in composition are observed. Different to the selected Upper Pliocene sediments, only the relative abundances vary, and not the type of constituents present. Therefore, the microfacies of the Lower Pliocene interval are more uniform and less clearly interpreted. Bioclast packstones (MF 12) and bioclast wackestones (MF 13) contain the same components (although large amounts are non-determinable biodetritus) and are mainly distinguished by the differences in the amount of sedimentary matrix. Bioclast-peloid packstones (MF 11) differ by higher amounts of fecal pellets. The differences in the amount of fine-grained matrix could reflect the varying input of fine-grained matrix from the platform top and thereby sea-level variations. This interpretation, however, is not unequivocal, because, during diagenesis, wackestones can be turned into packstones by compaction (see Chapter 5). This could account for some of the samples of MF 13. Both, sediments of MF 12 and 13 are moderately to well sorted and fine-grained and suggest that the material settled from suspension. Gradual shifts in the abundances of peloids and benthic foraminifers mark a slight cyclicity. These two groups of components could represent sea-level fluctuations by reflecting the changing distance of the locality from the shallow-water carbonate factory. During sea-level falls, the facies belts sourcing peloids and benthic foraminifers migrate down the gently dipping slope, and thereby move closer to the site of deposition, and retreat upslope during sea-level rises. A parallel increase in both groups would thus be expected during sea-level lowstands. Globigerinid packstones (MF 14) represent strongly marine sediments with low influence from the platform margin.

The globigerinid packstones (MF 14) represent the striking signal at the base of the Lower Pliocene interval examined (around 509 mbmp). These samples are dominated by planktic foraminifers and relatively high amounts of echinoderm fragments. The high amounts of planktic foraminifers decrease upwards within the lowermost meters of the interval examined. The lowermost Pliocene interval in CLINO below the selected Lower Pliocene interval extends over 30 m (536.33-507.70 mbmp), and has been described by Kenter et al. (in press) to show a similar predominance of planktic foraminifers and echinoderm fragments. These authors interpreted this pattern as an expression of a temporary drowning event in the earliest Pliocene. The Miocene platform was exposed by a pronounced sea-level drop at the Miocene-Pliocene transition, and subsequent flooding led to temporal drowning of the platform and the development of a marine hardground. The assumption of a drowning event is supported by the occurrence of phosphate at the hardground and in the pelagic sediments above (Kenter et al., in press). Similar phosphatized hardgrounds are described from drowning deposits in the Pleistocene succession of the Queensland Plateau (Glenn and Kronen, 1993) and from Jurassic drowning deposits of Jbel Bou Dahar, Morocco (Blomeier, 1997). Phosphate also occurs in the two basal Lower Pliocene samples examined here (see Chapter 5). Deposition above the hardground described by Kenter et al. (in press) was pelagic until shallow-water production was restored on the retreated platform that afterwards showed the gently dipping slope of a distally-steepened ramp. The fine-grained interval of the selected Lower Pliocene succession that is examined in the present study represents slope sediments of the recovered carbonate platform.

In the fine-grained sediments, that make up the majority of the selected Lower Pliocene interval, the fraction of skeletal grains (including the fine biodetritus and other not further determinable grains) varies around 50% of the sediment. The high amounts of fine-grained biodetritus and the strong diagenetic alteration of some components might be in part responsible for the monotonous appearance of the Lower Pliocene succession. The smaller grain-sizes of the biodetritus (compared to the N22 succession) also reflects the lower slope position of CLINO and its greater distance from the platform margin during the Lower Pliocene than in the Upper Pliocene. Throughout the selected Lower Pliocene interval, higher amounts of planktic foraminifers are observed than in the selected part of the Upper Pliocene succession (Fig. 16). The higher relative and absolute abundance of planktic foraminifers indicates a stronger open marine influence in the Lower Pliocene that at least partly is attributed to the lower slope setting. Additionally the low numbers of miliolids either point to a greater distance from the platform interior or they might result from the deeper and less protected environment on the top of the ramp in comparison with the flat-topped platform.

The relatively high amount of skeletal material in the selected interval of N19 from CLINO (approx. 50%) corresponds to the high abundance of skeletal material observed on the platform top of the Lower Pliocene pre-Lucayan Formation (to which foraminiferal zone N19 as a whole corresponds; Table 7;

Beach and Ginsburg, 1980). Reijmer et al. (1992) also observed a skeletal-rich composition in toe-of-slope turbidites in the Exuma Sound that are time-equivalent to the pre-Lucayan formation (Table 7). Components described from the platform interior of the pre-Lucayan ramp are small Miliolidae, agglutinating foraminifers, Amphisteginidae, *Halimeda*, red algae, echinoids, and corals (Beach, 1982). With the exception of corals, this association is also present in the CLINO samples from the studied N19 succession (mainly miliolids, echinoderms, and *Halimeda* plates).

	toe-of-slope		slope				platform interior	
reference	Reijmer et al., 1992		this study				Beach and Ginsburg, 1980	
deposits	pre-Lucayan	lower Lucayan	Lower Pliocene		Upper Pliocene		pre-Lucayan	lower Lucayan
			drowning	recovered	lowstand	highstand		
skeletal grains	●	○	●	●	○	+	●	+
non-skeletal grains	+	+	+	+	+	●	+	●
matrix [%] (platform derived)	45	75	39	53	20	63	PS	PS/GS
dominant compo-nents	n/a	n/a	non-determinable biodetritus, planktic, benthic foraminifers echinoderms	non-determinable biodetritus benthic foraminifers	coated grains, peloids non-determinable biodetritus, benthic forams, *Halimeda*	peloids benthic foraminifers non-determinable biodetritus	coralline algae, corals benthic forams bivalves (peloids), (ooids)	peloids, (ooids) mollusks benthic forams corals

Table 7. Comparison between time-equivalent deposits from the platform interior (Beach and Ginsburg, 1980), the slope (this study), and the toe of slope (Reijmer et al., 1992). Selected interval in the Lower Pliocene corresponds to pre-Lucayan Formation; selected interval in the Upper Pliocene corresponds to lower Lucayan Formation. To allow for a comparison of the data, the presentation has been slightly adjusted to the data from the present study. Black dots: dominant. Empty dots: abundant. Heavy crosses: present in considerable amounts. Light crosses: present in low amounts. PS: packstones. GS: Grainstones.

Beach and Ginsburg (1980) conclude from the composition that open circulation and a relatively great water depth on the bank itself, in combination with the absence of distinct marginal barriers, steered the development of these facies belts. Although the sediment composition observed in thin sections of the selected Lower Pliocene interval is in accordance to the findings of Beach and Ginsburg (1980), the palynomorph associations point to restricted conditions on the platform top. Large amounts of *Polysphaeridium zoharyi* indicate that elevated salinity prevailed in the source area. Foraminiferal assemblages, in contrast, exhibit no clear signature with respect to the environment they are sourced from (Fig. 18). The samples plot close to the corner of the triangle that represents 100% Rotaliina. This area is an overlap of different possible environments; hypersaline lagoons, normal marine lagoons, and normal open shelves. Although an interpretation can only be tentative, the possible shelf signature could reflect the absence of a strictly defined rim and the gently dipping slope morphology. Thus, the signature of palynomorphs and foraminifers could represent different facies belts of the platform at that time. Water exchange atop the platform thus seems to have been restricted despite the relatively open morphology. Laterally extensive banks, nevertheless, are prone to restriction even if devoid of margins, just by their large size hindering rapid water exchange (Simms, 1984). The dinoflagellate cyst assemblages that probably relate to a tropical and hypersaline paleo-environment are similar to the modern day western Great Bahama Bank. The lack of open oceanic and terrigenous influence in the palynomorph assemblages in the selected Lower Pliocene interval could indicate that westward offbank transport is the dominant mechanism for sediment export to the western slope with small to absent influence from Florida. This would agree with the seismic geometries as seen on the Western Line (Fig. 4) that show a strongly dominant progradation in a westward direction, implying a dominant sediment transport from the platform top to the western slope of the platform.

Numerical methods, based on component analysis data, indicate that there are no clearly separable statistical clusters that might represent typical highstand versus lowstand deposits. The slight fluctuations in the amount of peloids and benthic foraminifers show that there are traces of cyclicity but no significant shifts in constituents (Fig. 23). The more subtle variations in Lower Pliocene interval examined (except the aforementioned lowermost samples) are interpreted to reflect the gently dipping morphology of the platform that prevailed during the Lower Pliocene. Slope angles between the platform margin and CLINO are about 3.5°. The slope is thus steeper than those of a ramp in a strict sense (Ahr, 1973). The absence of well-marked highstand-lowstand differences could result from the gentle-sloping (ramp-type) morphology that prevailed during the Lower Pliocene (Beach and Ginsburg, 1980; Beach, 1982; Eberli and Ginsburg, 1987; Eberli and Ginsburg, 1989; McNeill, 1989). The effect of the distally-steepened ramp morphology, however, in the case presented here, is overlain by an overall long-term sea-level rise. The small scale sea-level fluctuations that have been examined are superimposed by a long-term sea-level rise, and could have resulted in a

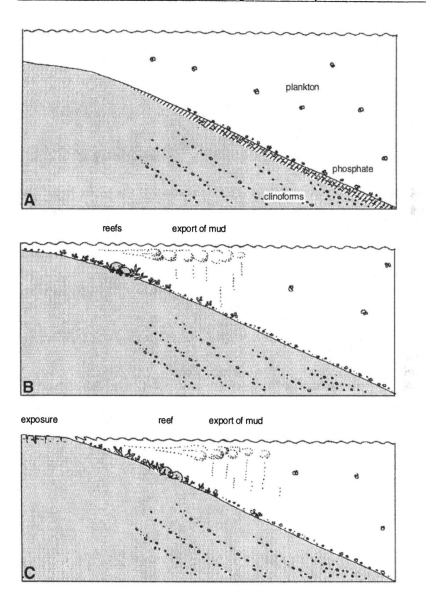

Fig. 23. Schematic sketch of the influence of sea-level fluctuations on the ramp-shaped platform in the Lower Pliocene. (A) The earliest Pliocene is characterized by a lower order drowning event with pelagic sedimentation and phosphatization of hardgrounds. (B) When carbonate production on the platform recovered, platform top-derived input reached the periplatform realm. (C) During higher order sea-level drops, facies belts shifted downslope and the productivity of the platform continued without drastic change.

stepwise sea-level rise rather than a succession of symmetric rises and falls. Sea-level fluctuations in the Lower Pliocene are recorded as infrequent exposure surfaces and facies shifts in sediments in the platform interior (Beach, 1982). A direct correlation of these observations to the slope sediments examined here, however, is not possible.

In summary, the apparent absence of clear compositional cyclicity can be attributed to a number of causes. First of all, the assumed high-frequency eustatic sea-level fluctuations are overlain by a long-term sea-level rise that caused the backstepping of the platform. Thereby, fast sea-level falls were presumably smoothed. Secondly, the interval sampled represents lower slope facies deposited at some distance from the shallow-water factory, where compositional variations are possibly less pronounced. Nevertheless, as has been shown by Everts (1994) and Everts and Reijmer (1995), compositional signatures of various platforms can be traced far out into the basin (Open shelf, Cretaceous: Everts, 1994; Everts and Reijmer, 1995; ramp, Mississippian: Elrick and Read, 1991). Therefore it is probable that in fact no clear signatures are present in the slope sediments of the ramp described. Presumably, key role can be ascribed to the gently dipping slope morphology of the Lower Pliocene carbonate platform. The shifts of facies belts on carbonate ramps that have been described to cause slight faunal changes as a response to sea-level fluctuations in the Cretaceous (Burchette and Wright, 1992; Bachmann et al., 1996; Bachmann and Willems, 1996) seem to be less pronounced to absent in the Lower Pliocene of CLINO. The slight cyclicity in the abundance of benthic foraminifers and peloids could possibly represent shifts of facies belts without faunal changes.

Clearly, more examples of ramps have to be investigated before a general rule can be extracted. Nevertheless, the distribution of peloids and foraminifers advocate that smaller scale sea-level changes resulted in a shift of the facies belts, and not in changes in the environments where the sediments are sourced from.

4.3.4
Sequence Stratigraphic Implications

The sea-level induced cyclicity observed in the selected Upper Pliocene interval exhibits the characteristics of highstand shedding. This is a smaller-scale repetition of the patterns observed in the large-scale cycles (Kenter et al., in press). These large-scale cycles are described to be composed of the same elements, coarse-grained lowstand deposits and fine-grained highstand deposits. In contrast to the larger-scale cycles of Kenter et al. (in press), the smaller-scale cycles examined within the present study seem to lack the clear drowning signature above the lowstand sediments (hardgrounds and phosphatization of the lowstand deposits). This possibly reflects the shorter time-spans involved in the high-frequency sea-level fluctuations. The mudstones found in the Upper Pliocene successions on top of the lowstand sediments could represent lag time deposits that would correspond to the temporary drowning of the larger scale cycles. A stacking of small-scale cycles seems to form the large-scale cycles.

The morphology of carbonate platforms (ramp versus rimmed) is one of the major controls on the response of a carbonate platform to sea-level changes. The higher-order sea-level fluctuations examined here especially influence sediment production on flat-topped platforms where, during sea-level lowstands, vast areas are exposed. On distally steepened ramps (in the wider sense used here), areas available for carbonate production are not reduced that drastically by the higher-order sea-level fluctuations, thus allowing for fairly continuous sediment production.

Sequence stratigraphic interpretations of sedimentary records of flat-topped carbonate platforms based on compositional analyses show the significance of sediment composition with respect to sea-level fluctuations (e.g. Haak and Schlager, 1989; Reijmer, 1991; Everts, 1994). Burchette and Wright (1992) proposed the application of stratigraphic analysis including facies successions to supplement sequence stratigraphic interpretations of classical ramps that, because of their large lateral extension and comparably low relief, frequently appear featureless. The study of facies successions in core or field sections is proposed as an additional tool for recognizing sequence architecture (Burchette and Wright, 1992). In some cases, clinoforms and sequence geometries of ramps have been successfully mapped using careful correlations of wireline, biostratigraphic and sedimentary logs (Stoakes, 1980; Cutler, 1983; Burchette and Britton, 1985; Chatellier, 1988; Dix, 1989). Bachmann et al. (1996) and Bachmann and Willems (1996) observed upward and downward shifts of the facies belts and thereby could prove the applicability of compositional analyses of ramp slope sediments.

Nevertheless, this study demonstrates that facies analysis of the periplatform sediments of a distally-steepened ramp might not yield a clear record of sea-level fluctuations. In response to a small relative sea-level fall, facies belts even on a distally steepened carbonate ramp are likely to shift basinwards. Whereas parts of the banktop will become exposed, production of carbonate sediment can continue on the former deeper slope. Therefore only slight changes in composition of the periplatform sediments of a ramp are likely to result from sea-level fluctuations.

4.4

Conclusions

The core CLINO offered the opportunity to investigate and compare slope sediments from a steep-sided, flat-topped carbonate platform (Upper Pliocene) and from a distally steepened ramp (Lower Pliocene). The focus has been on the signature of high-frequency sea-level fluctuations (4th order after Vail et al., 1991). The lower-order (3rd order) sea-level fall at the Miocene-Pliocene boundary where a subsequent rapid sea-level rise led to temporary drowning of the platform (compare Kenter et el., in press). When the shallow-water carbonate factory recovered, the platform retreated to form a distally steepened ramp. This ramp morphology developed into a steep-sided, flat-topped platform of the Upper Pliocene. The examination and comparison of sediments from the two distinct morphologic situations led to the following conclusions:

(1) The dependence of higher-order sea-level records in periplatform sediments on platform morphology has been demonstrated. Sea-level fluctuations that drastically affect the shallow-water factory of the interior of flat-topped platforms are clearly recorded in abrupt compositional changes. Lowstand deposits are characterized by coarse-grained sediments that lack significant amounts of fine-grained sedimentary matrix. Highstand deposits in contrast are characterized by a predominance of fine-grained material derived from the platform top. These observations once more verify the concept of highstand shedding (Droxler and Schlager, 1985).

In contrast, periplatform sediments of distally steepened carbonate ramps, where production zones rather migrate with the higher-order sea-level than being switched on and off, are relatively monotonous with only slight shifts in relative amounts of constituents. In the succession examined, however, an overall sea-level rise is superimposed on the high-frequency sea-level fluctuations, thereby possibly rendering the sea-level falls less pronounced.

(2) Beach and Ginsburg (1980) observed a change from a skeletal composition in the Lower Pliocene interval to a dominantly non-skeletal composition in the Upper Pliocene in platform interior sediments, and interpreted this change to reflect the morphologic evolution in the Pliocene. An analogous trend in composition was found in toe-of-slope turbidites (Reijmer et al., 1992). In the present study, a similar shift in composition was observed in the lower to upper slope sediments penetrated by CLINO.

(3) Palynologic information has turned out to be a valuable tool in interpreting sea-level history of a carbonate platform. Dinoflagellate cysts associations as well as pine pollen supplement petrographical observations by indicating environmental conditions such as salinity and temperature of the banktop waters. Also the existence of subaerially exposed land can be corroborated by pollen in a palynomorph assemblage.

(5) The aim of this part of the present study was to decipher the record of sea-level fluctuations with respect to the morphologic situation of a carbonate platform. Although diagenetic alterations rendered this task difficult, the methods applied possess the potential to answer the questions posed. Sampling for this study was relatively closely spaced in comparison with other studies in this field of research (e.g. Kenter et al., in press) that aimed to a larger-scale overview.

(6) In the past decade, research on climate focused on cyclicities of increasing frequency (Heinrich events, 7,000-10,000 y; e.g. Heinrich, 1988; Dansgaard-Oeschger cycles, 2,000 to 3,000 y; e.g. Dansgaard et al., 1982, 1993; Johnsen et al., 1992; Bond and Lotti, 1995). For a comparison with these studies that mostly deal with arctic or antarctic material, a tropical counterpart is found in periplatform carbonates, e.g. with stable isotopes. An additional control would be a high-resolution component analysis that also could expand the investigations to older material. Periplatform carbonates are characterized by comparably high sedimentation rates (e.g. in the intervals examined roughly 50 cm/ka and 15 cm/ka for Upper and Lower Pliocene, respectively) thereby possibly yielding the potential for high resolution studies.

5 Secondary Signals - Diagenesis

5.1

Carbonate Diagenesis

Previous work on the diagenesis of metastable carbonates has focused mainly on surface sediments from the platform top (for a review see James and Choquette, 1984). The Neogene sea-level fluctuations led to meteoric influence on most shallow-water sediments. Therefore, carbonate diagenesis has long been thought to be dominated by meteoric alterations. There is, however, a growing awareness that the conditions under which those sediments have been altered are not necessarily representative for the Earth's history, or for deeper-water settings (for an overview see Bathurst, 1993).

Periplatform carbonates are characterized by high fractions of bank-derived aragonite and high-Mg calcite, that are metastable in deep, cold sea-water environments (James and Choquette, 1983-b). This composition results in a high diagenetic potential of these slope sediments, in comparison to the lower diagenetic potential of pelagic, predominantly low-Mg calcitic carbonates, and of carbonates in tropical supersaturated marine waters (Schlager and James, 1978). Research on the diagenetic alterations of periplatform sediments is still in an early stage. Prior to 1985, when ODP Leg 101 took place, most information on carbonate diagenesis of platform slopes was drawn from piston cores from the Northern Bahamas (for reviews see Mullins, 1983, 1986). Studies of Schlager and James (1978), Saller (1984), and Mullins et al. (1985-a) belong to the few earlier investigations that were concerned with the deeper subsurface. In 1985, ODP Leg 101 offered the opportunity to study diagenetic alterations of periplatform sediments in deeper cores from the lower slope to toe-of-slope of the Bahamas (Dix and Mullins, 1988-a, 1988-b, 1992). The spatial link between the toe-of-slope to deeper slope sediments, and the platform top deposits was then provided by the cores of the Bahamas Drilling Project (Ginsburg, in press).

Nowadays, early near-surface marine-burial diagenesis, where normal sea-water causes dissolution, mineral stabilization and cementation, is considered an important process in lithification of marine carbonates and sandstones (Halley, 1987; Dix and Mullins, 1988-a; Raiswell, 1988; Bathurst, 1993; Maliva, 1995; Melim et al., 1995, in press-b; Hendry et al., 1996). For large portions of the

Bahamas Drilling Project cores, marine-burial diagenesis was confirmed by geochemical signatures (Melim et al., 1995, in press-b; Swart et al., in press-a, in press-b). Stable oxygen and carbon isotope compositions revealed that the upper 135 m of the CLINO core were influenced by meteoric fluids during early diagenesis (down to 153 mbmp; mud pit is 15 m above sea-floor; Melim et al., 1995). Between 135 and 153 mbmp, $\delta^{18}O$ and $\delta^{13}C$ values show a parallel downward shift towards higher values, thus implying a change in the diagenetic environment from meteoric to marine pore fluids (Melim et al., 1995, in press-b; Fig. 24). Therefore, the samples from CLINO below 153 mbmp offer a rare opportunity to examine Neogene shallow-water derived carbonates that have not been subject to meteoric influence.

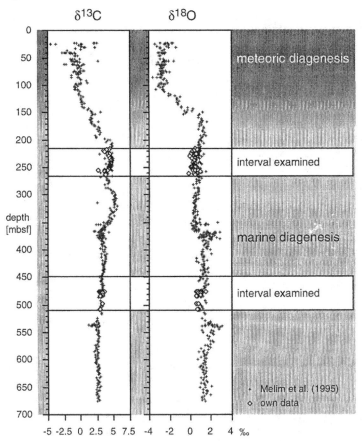

Fig. 24. Stable oxygen and carbon isotope compositions of bulk rock samples from the core CLINO, indicating an upper, meteorically influenced interval, and a lower interval that shows no indications of meteoric influence. Data from Melim et al. (1995, in press-b) kindly provided by L. Melim and P. Swart. Own data (rhombs) show the same trends.

A problem in reconstructing the processes of diagenesis is, that most studies are concerned with either recent soft sediments or ancient solid rocks (Schmoker and Halley, 1982). This concerns especially shallow-water derived carbonates in a deep-water realm. While recent low-latitude shallow-water carbonates mainly consist of metastable high-Mg calcite and aragonite, ancient carbonate rocks are composed of the more stable low-Mg calcite and/or dolomite (Bathurst, 1975). The transformation of unconsolidated metastable carbonates into lithified limestones with stable mineralogy involves a profound reorganization of components, matrix, and pores (e.g. Friedman, 1964; Gross, 1964; Land et al., 1967; Schlanger and Douglas, 1974; Steinen, 1978, 1982; Moshier, 1989). This generally hinders direct comparison of solid rock and soft sediment (Moshier, 1989).

The present study aims at a reconstruction of the early marine-burial diagenetic processes of periplatform carbonates. The Pliocene periplatform sediments of CLINO are well suited for this task since early diagenetic features have not been obscured by later diagenetic processes. Later diagenetic overprint is minor because neither deep-burial (in the order of km) nor tectonic deformation took place. In CLINO, the maximum depth of burial corresponds to the depth in the core, i.e. the samples investigated here have been overlain by less than 510 m of sediment.

The questions that are pursued here can be summarized as follows:
(1) How does diagenesis of shallow-water derived periplatform carbonates in a deeper-water environment proceed? (2) Are there diagenetic signatures that lead back to sea-level fluctuations and morphology evolution? (3) What kinds of diagenetic fluids have played a role in the diagenesis of the CLINO samples?

To approach these questions, first the diagenetic environment is examined with the aid of stable isotopes. Then the diagenetic alterations are described as investigated with X-ray diffraction, LECO (organic carbon), light microscope, and scanning electron microscopic. A lithification model for the periplatform carbonates is proposed. Finally, the signatures of fluid inclusions are discussed with respect to fluid flow.

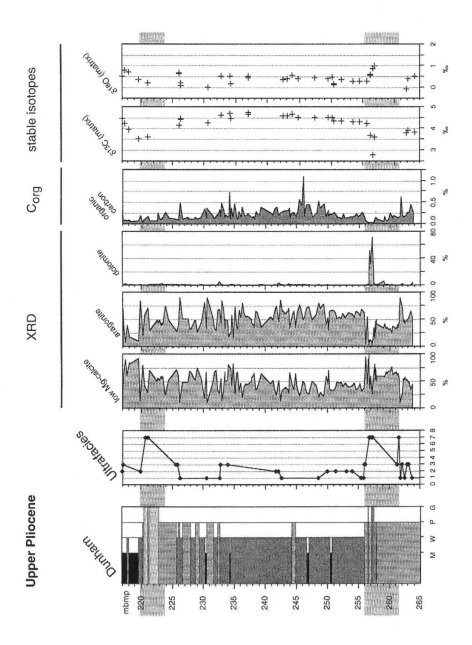

Fig. 25. Results of geochemical and mineralogical results of the selected Upper Pliocene interval. Also shown: Dunham classification from Chapter 4, and ultrafacies as described in Chapter 5.2.5.

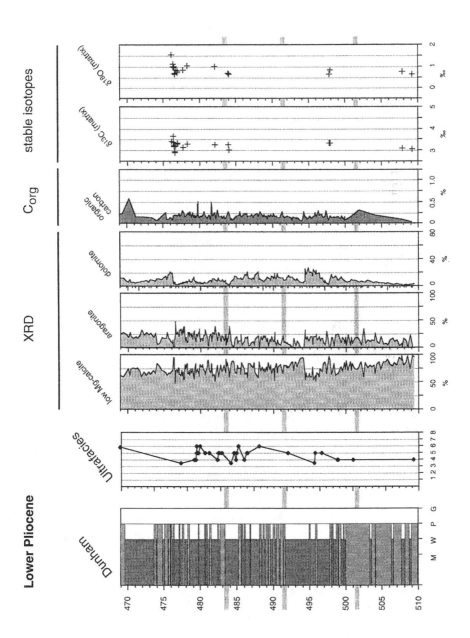

Fig. 26. Results of geochemical and mineralogical results of the selected Lower Pliocene interval. Also shown: Dunham classification from Chapter 4, and ultrafacies as described in Chapter 5.2.5.

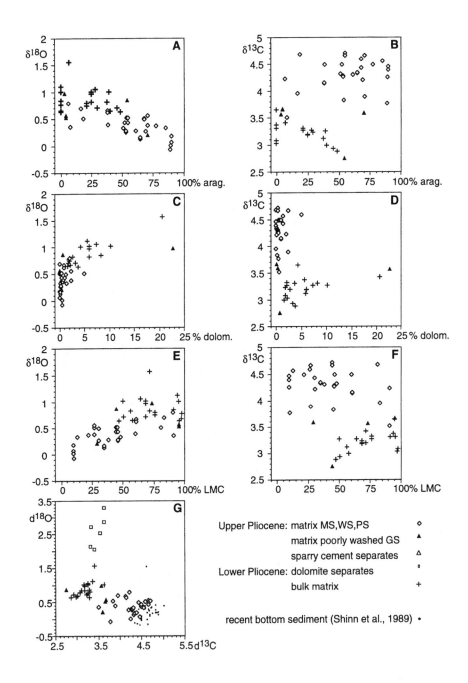

Fig. 27. Stable oxygen and carbon isotope compositions of the Upper and Lower Pliocene samples. (A) to (F): stable isotopes *versus* mineralogy. (G) Oxygen *versus* carbon. For comparison, the isotopic signatures of present-day platform top sediments taken from Shinn et al. (1989) are shown.

5.2

Results

5.2.1
Stable Isotopes

With the exception of one sample, the $\delta^{18}O$ composition of the bulk rock samples determined in the present study consistently shows positive values. Generally, the Lower Pliocene samples exhibit a more enriched composition than the Upper Pliocene samples (average $\delta^{18}O=0.87‰$, range 0.63‰ to 1.56‰ for selected Lower Pliocene interval; average $\delta^{18}O=0.41‰$, range -0.07‰ to 0.98‰ for selected Upper Pliocene interval; Figs. 25 and 26). Values for dolomitic subsamples are higher, reaching a maximum of 2.54‰ for a dolomitized component at 257.35 mbmp. These values are lower than those determined for pure dolomite separates by Melim et al. (in press-b).

As described by Melim et al. (in press-b) for their stable isotopes analysis of the entire core CLINO, the results presented here also imply that variations in the $\delta^{18}O$ composition correspond to variations in the carbonate mineralogy (Figs. 27). Low-Mg calcite concentrations co-vary with $\delta^{18}O$ values, i.e. diagenetically stabilized (low-Mg calcite-rich) samples exhibit heavier oxygen isotope signatures than the less altered aragonite rich samples.

The $\delta^{13}C$ composition of the Upper Pliocene samples exhibits higher values than the Lower Pliocene samples (average $\delta^{13}C=4.20‰$, range 2.75‰ to 4.71‰ for N22; average $\delta^{13}C=3.21‰$, range 2.88‰ to 3.64‰ for N19). Unlike the oxygen isotopes, no clear correlation between $\delta^{13}C$ and the mineralogy is observed.

5.2.2
X-Ray Diffraction

In XRD, three carbonate mineralogies (aragonite, low-Mg calcite, and dolomite) are observed (Fig. 28). Aragonite is present throughout most of the intervals examined. Because of its constant chemical composition and its therefore reliable position of the [111] peak (d=3.396 Å; datum from Joint Committee on Powder Diffraction Standards, 1970), aragonite was employed as standard mineral for peak corrections. In the samples examined, low-Mg calcite generally shows lower d-values for the [104] peak than expected for pure calcite (d = 3.027 ±0.002 Å, instead of d = 3.035 Å). This indicates that low amounts of $MgCO_3$ are present in the calcite, that result in decreased lattice distances (d-values; Goldsmith and Graf,

1958). The dolomite present exhibits d-values for the [104] peak of d = 2.905 ±0.002 Å. These values are somewhat higher than those expected of stoichiometric dolomite (d = 2.886 Å). The position of the [104] peak of dolomite is sensitive to stoichiometric variations. In dolomite, elementary cells are expanded by elevated amounts of $CaCO_3$ (Goldsmith and Graf, 1958). According to an equation of Lumsden (1979), this calculates to 43.7 mol-% $MgCO_3$ in the dolomite. Thereby, the dolomite present is a calcian dolomite. High-Mg calcite is absent from the intervals examined. In component analysis (Chapter 4), skeletal components were identified that are known to consist initially of high-Mg calcite, therefore the absence of high-Mg calcite seen in mineralogical analysis is a result of diagenesis.

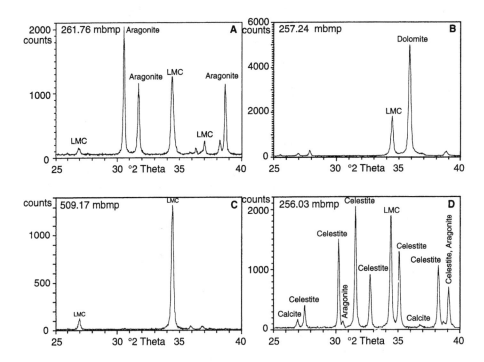

Fig. 28. Examples of X-ray diffractograms. (A) Sample with high amounts of aragonite and low-Mg calcite (LMC). (B) Sample with large dolomite peak and some low-Mg calcite. (C) Sample consisting almost exclusively of low-Mg calcite. (D) Sample with strong celestite peaks, and peaks of low-Mg calcite and aragonite.

In the Upper Pliocene, aragonite is an abundant mineral with an average of 47.8% of the total carbonate (Fig. 25). Aragonite varies strongly in relative abundance with minimum and maximum values of 1% and 90%, respectively. Aragonite shows a trend that roughly parallels the pattern of the fine-grained matrix as seen in component analysis. The lowstand deposits below 256 mbmp are characterized by aragonite values that first increase and then decrease upwards from 30% to 0%. An abrupt increase in the aragonite content to about 70% marks the base of the overlying highstand. This highstand is characterized by relatively high, but variable aragonite values that decrease towards the overlying lowstand where values again reach 10%. The selected Lower Pliocene succession shows less strong variations in the relative abundance of carbonate minerals (Fig. 26). Aragonite varies between 0% and 49% (average of 16.6%) of the total carbonate. Three faint cycles are observed with increasing values towards the tops.

Dolomite with an average of 2.8% in the selected Upper Pliocene succession shows a similarly variable abundance. Low amounts of dolomite are present throughout most of the Upper Pliocene succession. At the top of the lower lowstand in the selected Upper Pliocene interval (256.6-257.3 mbmp), dolomite reaches maximum values of 72%. High dolomite values, however, are rare, and only seven samples contain dolomite portions exceeding 10%. In the Lower Pliocene interval, dolomite averages at higher amounts (9.0%) than in the Upper Pliocene section, but shows lower maximum values. In the Lower Pliocene, dolomite varies between 2% and 28% and does not show similarly extreme deviations from the average as observed in the lowstand deposits of the selected Upper Pliocene interval. Dolomite contents show a complex pattern of increases and decreases with depth. These dolomite trends do not correlate to the vague cyclicity observed in component analysis, or to the aragonite cycles described above.

Low-Mg calcite is the most abundant carbonate mineral in both, the selected Upper and Lower Pliocene successions. In the Upper Pliocene interval, it amounts to an average of 49.4% of the total carbonate. Low-Mg calcite contents vary from 9% to 98%. With an average of 74.4% (maximum and minimum values of 99% and 48%, respectively), low-Mg calcite is by far the dominant carbonate mineral in the Lower Pliocene interval.

A scatter plot illustrates that aragonite and dolomite tend not to occur together in amounts exceeding about 25% (Fig. 29). This plot also shows a higher portion of aragonite and a lower portion of dolomite in the Upper Pliocene than in the Lower Pliocene, with the exception of some highly dolomitic samples in the Upper Pliocene.

Celestite infrequently occurs in samples of the Upper Pliocene as well as of the Lower Pliocene. In deeper marine deposits, celestite usually precipitates as nodules and is not disseminated in the sediment (Baker and Bloomer, 1988). Therefore, celestite usually is difficult to detect, and the occurrences observed cannot be quantified. Traces of quartz are present in some samples. Traces of quartz were distinguished in some samples, but are too close to the background to be quantified. Clay minerals form vague, broad peaks that amalgamate with the background and are too indistinct for further determination.

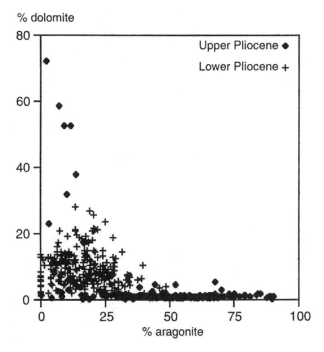

Fig. 29. Weight percent aragonite *versus* weight percent dolomite of the Upper Pliocene (rhombs) and the Lower Pliocene intervals examined (crosses). Samples with high aragonite portions usually contain low amounts of dolomite. The Upper Pliocene samples generally contain higher amounts of aragonite than the Lower Pliocene samples. An exception are some highly dolomitic Upper Pliocene samples, that contain low portions of aragonite.

5.2.3
Organic Carbon

Organic carbon is present in the selected intervals in low amounts that average at 0.22 weight % in the Upper Pliocene, and at 0.19 weight % in the Lower Pliocene (Figs. 25 and 26). The values measured thus plot at the lower end of values typical for carbonates (for a compilation see Ricken, 1993).

In both, the Upper and the Lower Pliocene intervals, organic carbon parallels the trend of aragonite contents, however, an offset between the two sets of samples is observed. (Fig. 30). In the Upper Pliocene, the lowstand interval at 256 mbmp is characterized by extremely low values of organic carbon, while the lowstand sediments at 222 mbmp are not clearly distinguished from the background. The content of organic carbon seems to be influenced by the amount of fine-grained matrix present, but also by the content of peloids (compare to Fig.

12). Parallel trends of organic carbon contents with the fine-grained matrix are not clearly seen in the Lower Pliocene, whereas a rough correlation is observed with the amount of peloids present (compare to Fig. 13).

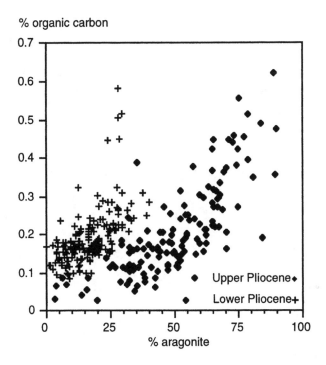

Fig. 30: Organic carbon *versus* aragonite contents measured in the selected Upper Pliocene (rhombs) and Lower Pliocene intervals (crosses). A positive linear relationship is observed for each of the two sets of samples.

5.2.4
Textural Alterations
SEM and Light-Microscopic Results

5.2.4.1
Diagenetic Alterations of the "Groundmass"

The "groundmass" (*sensu* Flügel, 1982; including sedimentary and diagenetic constituents) of the Pliocene periplatform carbonates is dominated by low-Mg calcite that occurs as micrite, microspar, and sparry cement, and by aragonite needles. The distinction between these different low-Mg calcite types is based on their grain size following the definitions of Folk (1959; 1965; 1974), the limits being defined at 4 μm between micrite and microspar, and at 30 μm between microspar and sparry cement. Folk (1959) described a gap in the distribution between microspar and micrite grains. Similar to observations of Munnecke (1997), no such gap was observed in the samples examined here (Fig. 31). Subordinately, dolomite, celestite, and phosphate are observed in the matrix. In the following, scanning electron microscopic (SEM) and light microscopic (LM) observations are described. For a description of diagenetic patterns on the basis of light microscopic examinations for the entire core CLINO, including the meteorically influenced upper part, the reader is referred to Melim et al. (in press-b).

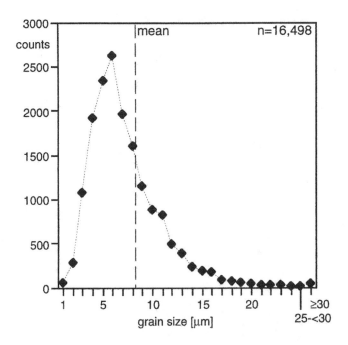

Fig. 31. Grain size frequency as determined on SEM micrographs (magnification 1000x). The largest apparent dimension visible of each grain was measured. Shown are the summarized counts of 46 samples.

Calcite.

Micrite: **SEM**: Crystallites ranging between <1 and 4 μm observed in the samples from CLINO exhibit irregular shapes (Plate 7A). The origin of the crystallites usually is not recognizable due to their small sizes. Micrite is found in low amounts in tightly cemented samples. In uncemented samples from the Lower Pliocene, micrite is a prominent constituent. **LM**: Uniform light brown colors to slightly laminated appearances characterize micrite-dominated samples under the light microscope. Inside foraminiferal tests (Plate 11A) and bryozoans (Plate 1G), peloidal micritic cements are rarely observed. Micritic rims of cortoids are common features in the coarse-grained sediments (see Chapter 4; Plate 4F).

Microspar. **SEM**: Microspar is the most ubiquitous "groundmass" constituent in the samples examined. Microspar crystals, proven as low-Mg calcite with the EDX, are characterized by equidimensional, anhedral shapes (Plate 7B). Amoeboidal microspar can form a tight mosaic. In porous samples, microspar also can exhibit subhedral shapes (Plate 7C). Towards larger components, microspar crystals exhibit sharp boundaries (Plate 7E). The size of the microspar crystals averages around 7.5 μm.

Two types of amoeboidal microspar can be distinguished. (1) In fine-grained deposits, microspar crystals enclose smaller components like aragonite needles (Plate 7B), or show characteristic pitted surfaces (Plate 7F). (2) In contrast, microspar crystals in coarse-grained lowstand deposits from the Upper Pliocene usually lack inclusions and pits. Similarly, inclusions as well as pits are absent from the microspar crystals where they infill closed fossil tests (Plate 9G). In geopetally filled fossil tests, the lower sediment-filled part of the interior is composed of microspar that includes sedimentary particles, whereas the upper part is free of inclusions (Plate 7G). In incomplete infills, microspar crystals are blocky-shaped to dog-tooth shaped. (3) In those Lower Pliocene samples that exhibit a high interparticle micro-porosity, a third type of microspar is found. In these samples, calcite crystals are common that reach sizes around 15 μm in longest apparent dimension. These crystals can exhibit subhedral shapes. They are usually not cut during the process of preparation but are rather removed from the relatively soft sample as a whole (Plate 7A), inhibiting the study of the internal structures (possible inclusions or pits). Smooth surfaces, however, point to a probably absence of inclusions and pits.

LM: Similar to the SEM examinations, where inclusion-rich and virtually inclusion-free microspar can be distinguished, two appearances of tight microspar mosaics are observed under the light microscope. (1) Microspar in the fine-grained sedimentary matrix appears light-brown, similar to the micrite matrix, from which it is difficult to distinguish. (2) Clear microspar is observed in coarse-grained samples and inside fossil tests. Isopachous linings of microspar or incomplete infills are common in fossil tests. (3) In some Lower Pliocene samples that show a high intercrystalline porosity, microspar crystals exhibit subhedral shapes.

Sparry cement. **SEM**: Sparry cement crystals with grain sizes greater than 30 μm are rare. They are observed as interparticle cement in coarse-grained samples that lack or contain only low amounts of sedimentary matrix, and as intraparticle cement in fossil tests (Plate 7H). When constituting interparticle cement, the spar crystals are equidimensional and anhedral (Plate 12G). These sparry cements usually contain no inclusions or pits. Sparite cements occur together with the (virtually) inclusion-free type of microspar in the lowstand deposits. There is a continuous transition between sparry cements and inclusion-free microspar. As internal cements in fossil tests and in secondary pores, shapes vary from blocky to elongated spar. Inclusions are absent from the internal sparry cements. *LM*: Clear sparry cements characterize the coarse-grained lowstand deposits (Orthospar *sensu* Wolf and Conolly, 1965). In these samples, sparry cements occur together with clear microspar that has a similar appearance. In shelter porosity in coarse-grained samples, elongated, scalenohedral sparite reaches 100 μm in length. Inside incompletely cemented fossils, isopachous palisades of blocky cement are common (Plate 11B). Neomorphous sparite in e.g. mollusk shells (Pseudospar of Folk, 1965) sometimes is distinguished from a primary precipitate by ghost structures (Plate 1E; see Chapter 5.2.4.2). Sparry cement commonly infills leached cortoids where only the micritized rim is preserved, with non-isopachous linings to equant spar cements (Plate 5A). Under the cathode beam, no luminescence is observed in these sparry cements (or any other diagenetic precipitates). This corresponds to the behavior of most post-Miocene carbonates (Major, 1991).

Syntaxial Overgrowths. **SEM**: Syntaxial overgrowths around bioclastic grains are common in a wide range of sizes. Syntaxial rims are found to grow mainly on foraminifera (especially globigerinids, Plate 7H) where they occur mainly inside the test but also grow outward. Dog-tooth to elongated scalenohedral shapes characterize internal syntaxial overgrowths in foraminifers. Syntaxial overgrowths are also common biodetritus (Plate 7D) such as echinoderms and more rarely on bryozoans. On echinoderms, blocky crystals continue in the direction of the monocrystalline skeleton. *LM*: Under the light microscope, syntaxial overgrowth is a conspicuous pattern especially in coarser-grained sediments. Optically continuous, syntaxial rim cements that surround echinoderm fragments often show larger grain sizes than the adjacent crystals, implying a rapid growth of the syntaxial overgrowth (Plate 11D). Sweeping extinction is typical for syntaxial overgrowths on foraminifers (mostly globigerinids but also rotaliids).

Aragonite.
Aragonite needles. **SEM**: Aragonite needles that compose a large portion of the sedimentary matrix were described before (Chapter 4). They appear degraded by dissolution (Plate 4H).

Acicular cements. **SEM**: Acicular cements are extremely rare, and have been observed as internal cements only in three fossil tests (at 261.82 and 262.18 mbmp). These internal needle-shaped cements, that are 1-2 μm thick and up to 15 μm long, are subsequently cemented by blocky cement and thus predate these cements (Plate 8A). The mineralogy of the needles could not be determined, the occurrence of pitted internal blocky cements in one sample (476.40 mbmp, not shown) indicates a possibly aragonitic original mineralogy of the needles. Morphologically similar acicular cements enclosed in a second generation of blocky spar have been described by Sandberg (1985). *LM*: In thin section, needle cements are also extremely rare and are found as intraskeletal linings around *Halimeda* utricles in one strongly neomorphous specimen (Plate 11E; see below).

Dolomite.

Different types of dolomite are observed that have in common a calcian non-stoichiometric composition and correspond to protodolomites *sensu* Graf and Goldsmith (1956). They contain around 43 mol-% of $MgCO_3$ as measured by EDX and calculated from the shift of the position of the [104] peak in XRD analysis (Chapter 5.2.2).

Dolomite rhombs. SEM: Single euhedral dolomite rhombs, ranging from 1 to 20 μm in length, are present in the matrix throughout the intervals examined. The rhombs are usually well developed with sharp and straight edges (Plate 8B). Rarely, dolomite rhombs appear pitted. Dolomite rhombs also occur as primary and as secondary pore fills, and in Lower Pliocene samples are also rarely found inside pellets. Dolomite crystals smaller than 1 μm are present included in echinoderm fragments (Plate 9F; see below). Similar micro-dolomite inclusions have been described from various diagenetically altered, originally high-Mg calcitic biota (Rush and Chafetz, 1991). - *LM*: The most striking type of dolomite seen in thin section is the rhombohedral secondary cement in leached cortoids.

Microcrystalline dolomite. SEM: Dolomite also occurs as microcrystalline (<1μm), dense to sucrosic matrix (Plate 8C, D). In this microcrystalline matrix, no aragonite needles are engulfed. Components are surrounded by the dolomite matrix, but are usually still preserved as calcite. In a number of samples, subhedral rhomboidal to cryptocrystalline dolomite forms envelopes around sedimentary components. These envelopes appear as primary cements that fill in wall pores in foraminiferal tests (Plate 8C, D) and endolithic borings (Plate 8E) and clearly are non-destructive. - *LM*: In samples, where the matrix is dolomitic, primary structures of the bioclasts are well preserved.

Celestite ($SrSO_4$). SEM: Celestite is a rare constituent in the samples examined. It occurs as matrix (Plate 8F), and subordinately in neomorphosed bioclasts (Plate 8G). Individual celestite crystals are usually not distinguishable because, unlike carbonate minerals, the margins of celestite crystals are not accentuated by the etching process (Plate 8F). Frequently, celestite surfaces are pitted or include molds (Plate 8F). Where they bound molds, the edges of the celestite precipitates are sharp. The assessment of the amount of celestite in a sample is difficult because celestite occurs locally as nodules while it might be entirely absent from the remaining volume of a sample. *LM*: In light microscope, celestite is inconspicuous. Sometimes brownish colors of bioclasts indicate the presence of celestite.

Pyrite (FeS$_2$). *SEM* and *LM*: Pyrite is rare in the samples examined. It exclusively is present as framboids in the order of 15 µm in diameter. These framboids mostly are found inside foraminifer tests (Plate 11F) and rarely in the fine-grained matrix.

Phosphate (Ca$_5$(PO$_4$)$_3$).— SEM: Phosphate occurs only in one Lower Pliocene sample (509.17 mbmp), although it is reported from other parts of core CLINO, for example from a major hardground below the Lower Pliocene succession examined (536.33 mbmp; Melim et al., in press-b). Tiny (< 1µm) rough-surfaced crystals compose part of the matrix that bounds the larger components without destroying them (Plate 8H). Similar to dolomite, phosphate also forms cryptocrystalline envelopes around components.

5.2.4.2
Diagenetic Alterations of Sedimentary Components

Diagenetic alterations are not only documented in the "groundmass" constituents described above, but also in (mostly biogenic) larger grains. Metastable high-Mg calcitic components are either structurally rather well preserved but show a lowered Mg-content of the skeletal calcite, or they are dissolved leaving molds. Similarly, aragonitic components are subject to dissolution and cementation.

Red algae. SEM: In SEM, corallinaceans are characterized by dolomitized walls around the individual cells (Plate 9A and B). The two-layer-structure of corallinaceans described by Flajs (1977) is clearly discernible. The "secondary" layer that encloses the cell and originally is composed of high-Mg calcite, is dolomitized. The "primary" layer that divides the "secondary" layers, and initially is either also of calcitic (*Lithothamnion*-type) or of organic material (*Goniolithon*-type), appears as a gap or is also dolomitized. To which of the two types the specimens present originally belonged could not be decided, because the high-Mg calcitic "primary" layer of the *Lithothamnion*-type is initially composed of needles that could be dissolved whereas at the same time the prismatic secondary layer could be preserved. *LM*: In low magnifications, red algae appear well preserved (Plate 1B and C), in higher magnifications, however, micritic recrystallisation (dolomite) obscures the microstructures.

Halimeda. SEM: Low-Mg calcite crystals are found to internally cement *Halimeda* fragments by enclosing their aragonite needle skeleton (Plate 9D) similar to the microspar cement in the matrix that engulfs aragonite needles. Calcitization of a variety of components by this type of cementation was described by Sandberg (1984, 1985) and Sandberg and Hudson (1983). This cementation of *Halimeda* fragments is not always complete, sometimes it is restricted to the tubes, leaving the inter-tube space as cavities (Plate 9C). In some of the Pliocene samples from CLINO, celestite is found, instead of calcite, to enclose the needle skeleton (Plate 8G shows an overview of a celestitized specimen). *LM*: *Halimeda* fragments are frequently neomorphous, i.e., they are preserved as blocky cements that yield ghosts of the primary vesicles (Plate 11E). As noted by Melim et al. (in press-b), these neomorphous fragments often show a characteristic pseudopleochroism. This property is thought to originate from inclusions (Hudson, 1962). As seen in SEM, *Halimeda* fragments contain numerous aragonite needles that might cause the apparent absorption. *Halimeda* plates also can be preserved as clear sparry cements with micritic rims in the place of the former tubes (Plate 11G). Due to the metastable primary mineralogy, *Halimeda* plates also are frequently dissolved leaving molds.

Mollusks. SEM: Mollusks are rarely observed under the SEM. Whereas calcitic shells (bivalves) are usually preserved, aragonitic mollusks (gastropods) are frequently either dissolved leaving molds, or are entirely replaced by microspar (Plate 7G). *LM*: Most gastropods, being typified by an aragonitic initial mineralogy, are dissolved leaving molds. Molds of larger fragments are still recognizable as gastropods, while molds of smaller-sized detritus is indeterminable. The molds are frequently hollow, but secondary cementation that fills in the former mold also occurs (Plate 1E). In those cases, no skeletal structures are preserved. Less frequently, fragments are found that still bear ghosts of primary structures, and similar to neomorphous *Halimeda* fragments show pseudopleochroic extinction. Bivalve tests are usually well preserved and still show their lamellar structure. Bioerosion (borings) is infrequently observed in bivalve shells (Plate 1H).

Echinoderms. SEM: Echinoderm fragments are preserved fabric-retentively. Nevertheless, unmixing of the originally metastable high-Mg calcitic echinoderm fragments in stable low-Mg calcite is recognized by micro-dolomite inclusions in the now low-Mg calcitic fragments (Plate 9E and F). The presence of micro-dolomite inclusions in low-Mg calcite is thought to be a strong indicator of an originally high-Mg calcitic mineralogy (Lohmann and Meyers, 1977) and was frequently reported from fossil echinoderms (e.g. Rush and Chafetz, 1990). The microscopically visible skeletal structure of the fragments remains unaltered by this process. Also, syntaxial cementation of intraskeletal pores is observed. *LM*: Echinoderm fragments are characterized by a clear appearance and uniform extinction under crossed Nicols. In deeper parts of the core, echinoderm fragments can be internally replaced by micrite-sized dolomite crystals, thus showing a less clear single-crystal extinction pattern (Plate 11D).

Foraminifers. SEM: The microstructures of planktic foraminifers that are composed of low-Mg calcite are mostly well preserved. The preservation of low-Mg calcitic benthic foraminifers, in contrast, varies. Amphisteginids still show their original microstructure. Preservation of rotaliids varies between excellent conservation of the wall structures (layers and pores) and recrystallization. Remains of the inner organic layers (linings) are frequently observed (Plate 9H, see also Plate 2B). Initially high-Mg calcitic miliolids are usually dissolved or replaced (Plate 9G). In some cases, nevertheless, needles of 1-3 μm in length that initially compose the miliolid tests are preserved. *LM*: Planktic as well as rotaliid foraminifers are characterized by a clear appearance (Plate 11A). Crushed Globigerinids are observed in strongly compacted samples (Plate 11H). Frequently, the inner organic layer of Rotaliids, that typically appear reddish-brown under the light microscope, is preserved. Miliolids with thick tests are usually well preserved exhibiting their characteristic brown, almost opaque color (Plate 11C). Thin-shelled specimens, in contrast, are usually dissolved, frequently after internal cementation, leaving delicate molds that later can be infilled by secondary cements (Plate 3A). This renders recognition of these miliolids progressively difficult and could lead to a strong reduction in the fossil record.

Fecal pellets. SEM: Even in tightly cemented samples, pellets often appear internally unaltered, still consisting of aragonite needles (Plate 4D; aragonitic mineralogy was proven using Feigels solution under the SEM using the method of Schneidermann and Sandberg, 1971). Pellets are known to resist diagenetic stabilization of aragonite longer than other aragonitic components such as *Halimeda* fragments (Beach, 1982). Pellets often show slightly cemented rims that are comparable to those found in recent pellets from the modern platform top of the Bahamas. These rims could have influenced the preservation of the aragonite needles. *LM*: Pellets are characterized by a brown color. They are present in various states of deformation, from round or elliptical in cemented samples to strongly deformed or homogenized in compacted samples. In some samples, the interior of the pellets is dissolved.

Palynomorphs. SEM: Most SEM samples exhibit up to five dinoflagellate cysts. Samples that appear compacted as indicated by deformed peloids and diagenetic laminations, generally contain considerably higher amounts of palynomorphs than uncompacted layers (up to 20 specimens). Wall structures are usually well preserved, but the deformation of the initially spherical vesicle differs conspicuously. For the most common dinoflagellate cysts, that are characterized by a spherical vesicle of 20 to 100 μm in diameter with a thin organic wall (less than 1 μm), and delicate spines of about 10 μm in length (see Chapter 4), deformation is easily assessable. Two general styles of preservation of thin-walled organic microfossils can be distinguished: (1) In uncompacted layers, dinoflagellate cysts frequently exhibit their original spherical to slightly deformed shape, whereas completely flattened specimens are absent. Despite of their very thin walls, the spherical shape of the cysts can be preserved at least down to the base of the interval examined (510 mbmp). Most of the dinoflagellate cysts are hollow (Plate 10A), some are filled with carbonate cement (Plate 10B). Spines of most of the spherical specimens are preserved in their original orientation directing radially away from the bodies. (2) In compacted layers, dinoflagellate cysts are always flattened (Plate 10C and D), and usually oriented parallel to the bedding. Spherical specimens do not occur.

5.2.5
Description of Ultrafacies

Scanning electron microscopic observations led to the recognition of distinct types of textures present in the Pliocene periplatform carbonates. These textures are characterized by variations in the relative amounts of the different "groundmass" constituents, namely aragonite needles, micrite, microspar, sparite, and dolomite. Following the suggestion of Keupp (1977), the distinct texture-types are termed "ultrafacies". Analogous to the term "microfacies" of Flügel (1978) as used in Chapter 4 of the present study, Keupp (1977) proposed the term "ultrafacies" to describe the petrographic and paleontologic characteristics of sediments, below the limit of resolution of a light microscope. In the present study, the emphasis in the description of the ultrafacies is laid on the diagenetic features with minor focus on the paleontologic composition.

In the following, first the diagenetic patterns of the fine-grained sediments (highstand deposits of the Upper Pliocene and the Lower Pliocene deposits) are described, then the coarse-grained Upper Pliocene lowstand deposits are presented.

5.2.5.1
SEM Observations in Samples with Fine-Grained Matrix

The fine-grained deposits of the Pliocene samples exhibits six distinct textures and combinations of the above described matrix constituents. The first three ultrafacies are predominantly found in the Upper Pliocene succession, and the fourth to sixth exclusively in the Lower Pliocene succession (Figs. 25, 26):

Ultrafacies 1. Aragonite needle mesh (n=11). Ultrafacies 1 (Plate 12A) consists of carbonate muds that are composed of an irregular mesh of needles, and low amounts of tiny crystallites (< 2 μm). The needles are of aragonitic mineralogy as is indicated by correlation of ultrafacies 1 with aragonite-rich samples identified in XRD analyses (Fig. 25). The needles reach lengths of up to 10 μm, but most are

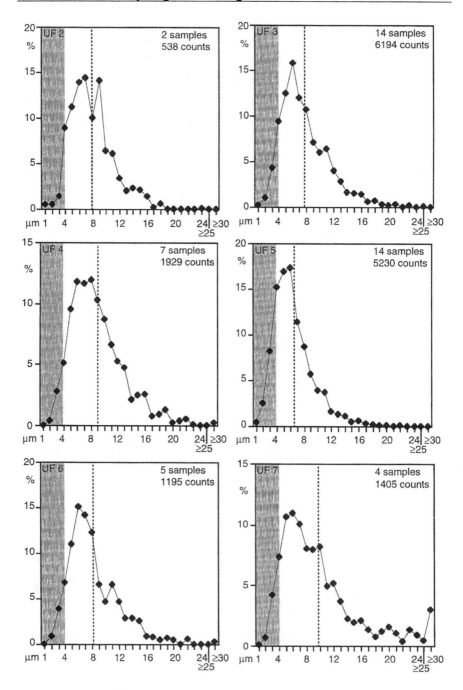

Fig. 32. Frequency distribution of the grain sizes of cement crystals and crystallites in the matrix. Shaded area between 0 and 4 µm represents the range of micrite. Dotted line marks the mean. Counts are normalized to percent for each sample prior to summarizing.

smaller (3-6 μm; Plate 4H). Unetched samples reveal that the needles appear degraded. They frequently show pointed terminations and poorly developed crystal surfaces and thus resemble aragonite needles from whitings as described by Macintyre and Reid (1992) (see description in Chapter 4). Micro-dolomite rhombs of 0.5-2 μm in the longest apparent extension are dispersed in the needle mesh (Plate 12A). Micro-dolomite rhombs are also present enclosed in echinoid fragments that, originally being composed of high-Mg calcite, are preserved as low-Mg calcite crystals (Plate 9E and F). Benthic as well as planktic foraminifers are internally hollow to entirely cemented. Rare pyrite is found inside some foraminifers. Thickness reduction is indicated by strongly deformed organic-walled microfossils, and by broken foraminiferal tests as observed under the light microscope (Plate 11H). Mechanical porosity reduction by an overburden exceeding 200 m of sediment nevertheless left a high interstitial porosity (Plate 4H).

Ultrafacies 2. Partially microsparitic needle mesh (n=7). Several samples contain microspar crystals of up to 10 μm in diameter (average around 7.9 μm, standard deviation 3.1 μm; Fig. 32) that are scattered throughout the loose aragonite needle matrix (Plate 12B). These microspar crystals enclose aragonite needles and other small components and mostly show a rounded shape. Some microspar crystals, however, exhibit a roughly euhedral shape. Several single microspar crystals are composed of a biogenic center as, for example, an echinoderm fragment and a cement rim which engulfs aragonite needles (Plate 7D). Foraminifers are usually internally filled by cement crystals, but infrequently empty tests are observed.

Samples exhibiting such moderate amounts of microspar crystals still possess a high intercrystalline porosity. Dinoflagellate cysts are mostly flattened. Pores of apparently secondary origin ("micro-vugs" of Moshier, 1989) are common, which are 10 to 30 μm in diameter.

Ultrafacies 3. Microsparitic limestone (n=18). While the first two ultrafacies are present only in the Upper Pliocene samples, ultrafacies 3 is present in both, the Upper and the Lower Pliocene successions. A dense mosaic of anhedral microspar crystals with slightly curvilinear boundaries ("amoeboidal" *sensu* Fischer et al., 1967) exhibit a considerably reduced intercrystalline pore space. Small components, especially aragonite needles, are entirely embedded in the microspar crystals, that correspond to the first type of the amoeboidal microspar described in Chapter 5.2.4.1 (Plate 12C). The microspar crystals show a range of grain sizes from 3 to 30 μm (mostly 6 to 12 μm; average 7.8 μm, std. dev. 3.4 μm; Fig. 32). Bioclasts that are initially composed of low-Mg calcite possess sharp boundaries against the enclosing microspar (e.g. ostracods; Plate 7E) and generally are well preserved showing their original ultrastructure. Benthic as well as planktic foraminifers in ultrafacies 3 are entirely filled with cement that sometimes includes tiny pyrite framboids. The frequent molds usually show secondary cementation. Thin-walled organic microfossils are spherically preserved or slightly deformed. The delicate spines of dinoflagellate cysts are usually embedded in the cement crystals pointing radially away from the body. Peloids (fecal pellets), consisting of aragonite needles, are well preserved. They are tightly surrounded by microspar crystals that do not extend into the interior of the pellets. Components initially consisting of high-Mg calcite show two types of preservation. Echinoderms are transformed into low-Mg calcite crystals that bear tiny enclosed rhombohedral micro-dolomites (< 1 μm; Plate 9E and F). The high-Mg calcitic walls of miliolid foraminifers, in contrast, are frequently dissolved. Rotaliid tests vary in preservation from largely unaltered to entirely recrystallized.

Celestite occurs in several entirely cemented samples of both Upper and Lower Pliocene (256.18, 262.18, 474.09 mbmp). Where occluding the interparticle pore space, celestite engulfs aragonite needles or appears pitted (Plate 8F), thereby implying that it precipitated as an early cement. In one sample (256.03 mbmp), *Halimeda* fragments are cemented by celestite (Plate 8G). The aragonite needles, which form the calcareous skeleton of the algae, are preserved being engulfed in the celestite cements.

Ultrafacies 4. Pitted microspar (n=10). In the Lower Pliocene interval, lithified samples can exhibit a dense mosaic of microspar crystals averaging at 8.9 μm in diameter (std. dev. 3.7 μm; Fig. 32). The microspar crystals in ultrafacies 4 are characterized by empty pit structures (Plate 12D). In these pitted microspar crystals, no enclosed aragonite needles are present (Plate 7F). Celestite has not been observed. Otherwise the matrix of ultrafacies 4 closely resembles that of ultrafacies 3.

The interior of foraminifers usually is tightly cemented, and dissolution molds contain secondary cements. Some peloids show diffuse boundaries and marginal cementation. They often contain dolomite rhombs, and aragonite needles are less abundant. Similar to ultrafacies 3, dinoflagellate cysts in ultrafacies 4 are preserved spherically to slightly deformed.

One single Lower Pliocene sample (509.17 mbmp) that otherwise shows the characteristics of ultrafacies 4, deviates from the mineralogical spectrum described so far. Here, rare phosphate crusts around components are observed, and phosphate also locally occurs as matrix (as measured by EDX; Plate 8H). This sample already was conspicuous in component analysis by the high amounts of planktic foraminifers that point to low sedimentation rates (Chapter 4).

Ultrafacies 5. Compacted carbonate (n=17). Alternating with ultrafacies 3 and 4, samples occur in the Lower Pliocene succession, that are characterized by high interparticle micro-porosities and smaller grain sizes with an average diameter of 6.5 μm (std. dev. 2.8 μm; Fig. 32). They consist predominantly of small crystallites, apparently composed of low-Mg calcite, and of larger, roughly subhedral calcite crystals of microspar size. Amoeboidal microspar, in contrast, is virtually absent. Dolomite rhombs, that reach lengths of up to 15 μm in the largest apparent dimension, are present in varying amounts. Aragonite needles are rare and poorly preserved. The interior of benthic and planktic foraminifers varies from hollow to entirely cemented tests, but most tests are partly filled with cement. Molds and micro-vugs in the order of 10-50 µm in diameter lack secondary cementation. Rarely, local celestite precipitates occur.

Under the SEM, as well as under the light microscope, signs of compaction are observed, such as deformed peloids, deformed burrows, and flattened organic microfossils (dinoflagellate cysts).

Ultrafacies 6. Transitional (n=4). Transitional between ultrafacies 4 and 5, some Lower Pliocene samples exhibit partial cementation. They consist of small calcite crystallites and larger subhedral calcite crystals like ultrafacies 5, but locally are dominated by amoeboidal microspar like ultrafacies 4. The average size of the grains present is 8.2 µm (std. dev. 3.6 µm; Fig. 32). Foraminiferal tests are partly to entirely cemented. Various degrees of deformation of organic-walled microfossils are observed. As estimated on the basis of SEM micrographs, interstitial porosity is slightly lower than that of ultrafacies 5.

The six ultrafacies do not occur in a clear depth-dependent order but are rather found in irregular alternations. An overall downcore trend is observed from ultrafacies 1 (loose needle mesh), ultrafacies 2 (partially microsparitic needle mesh), and ultrafacies 3 (tight microspar that encloses aragonite needles) in the Upper Pliocene, towards ultrafacies 4 (pitted microspar crystals), ultrafacies 5 (compacted samples with small calcite crystallites), and ultrafacies 6 (transition between the two latter) in the Lower Pliocene.

5.2.5.2
SEM Observations in Grainstones

Ultrafacies 7. Cemented limestones (n=5). Coarse-grained lowstand deposits of the Upper Pliocene succession are characterized by low intercrystalline porosity (Plate 12G). Sparry cements occur together with microspar that usually lacks inclusions. This inclusion-free microspar shows a morphology similar to the sparry cements, and is distinguished only by its smaller size. Grain sizes of the calcite constituents vary significantly from 1 to 41 µm and average at 9.9 µm (std. dev. 4.7 µm; Fig. 32). In some samples, low amounts of aragonite needles or pitted structures are observed enclosed in the sparry cement. Components are usually sharply bounded by sparry cement and microspar crystals. Epitaxial overgrowth is common on both, planktic and benthic foraminifers. Internal, as well as external cements are observed to follow the crystallographic direction of the test crystallites.

Dolomite rhombs are disseminated in the matrix in various amounts. Rarely, fine-grained dolomite (1 µm) locally replaces the matrix. Thin crusts of cryptocrystalline dolomite (<1 µm) frequently cover components and occlude the pores, for example, of benthic foraminifers. Several components can share one delicate dolomite fringe. Within individual samples, different cement successions can be found. Dolomite as primary fringe on a component, can be followed by a calcite cement seam, while around other components, a primary calcite fringe is followed by a dolomite envelope. Calcite cements in direct contact with dolomite envelopes sometimes show an elongated, bladed shape that resembles the morphology of high-Mg calcite cements. EDX analyses, however, reveal the low-Mg calcite nature of those crystals. Although no unequivocal micro-dolomite inclusions were found, it cannot be excluded that these cement fringes originally precipitated as high-Mg calcite. Celestite is rarely found to occlude the interparticle pore space in the sparry cemented samples (261.57 mbmp).

Microstructures of components initially consisting of low-Mg calcite (e.g. ostracods, planktic foraminifers) are well preserved. These shells are sharply bounded by cement crystals. Molds are common that are thought to originate from initially metastable bioclasts. Molds are frequently partly infilled by secondary calcite and dolomite cements. Peloids, consisting predominantly of aragonite needles, are well preserved. They usually still consist of an aragonite needle mesh and show well preserved, uncemented, burrows. This fresh appearance is conspicuous considering the occurrence of molds in the same samples.

The sparry cemented coarse-grained samples appear undeformed. Peloids are preserved in an oval shape that is thought to be the original appearance. The rare dinoflagellate cysts are preserved spherically with their spines pointing away from the bodies.

Ultrafacies 8. Dolomitized grainstones (n=3). Some coarse-grained samples in the lowstand deposits are strongly dolomitized (256.79, 256.82, 257.24 mbmp; Plate 12H). In these samples, the matrix consists of fine-grained dolomite crystals (usually around 1μm, but reaching up to 20 μm in the longest apparent dimension), and local inclusion-free microspar mosaics. Foraminifers are internally cemented, some by tight calcite cements, others by dolomite rhombs, and some by both. As seen in EDX analyses, the dolomite is of calcian composition. Larger components are still preserved as low-Mg calcite, but their boundaries sometimes appear affected, thus implying that the dolomitization might be at least partly replacive. Other components are well preserved and are covered by a dolomite crust as described to occur in ultrafacies 7.

Moldic pores with diameters around 50 μm are observed. Like in ultrafacies 7, the samples with ultrafacies 8 characteristics appear uncompacted as indicated by apparently undeformed peloids and dinoflagellate cysts.

5.2.6
Fluid Inclusions

In the coarse-grained Upper Pliocene lowstand deposits, fluid inclusions were investigated microthermometrically in order to gain information on the diagenetic fluids that contributed to diagenesis. Because sparry cement crystals are required for the examination of fluids included in the cements, only data from the lowstand deposits of the Upper Pliocene could be obtained. Two samples were analyzed that exhibit sparry cements with inclusions large enough for analysis (220.49 and 257.35 mbmp). In 220.49 mbmp, inclusions in the clear overgrowth on an echinoderm fragment were analyzed, whereas in 257.35 mbmp, inclusions in cloudy bladed cements were examined.

In the samples examined, exclusively all-liquid fluid inclusions are present. This is typical for low-temperature phreatic diagenetic environments (Goldstein and Reynolds, 1994). These fluid inclusions have been recognized as primary inclusions. Ice melting temperatures (Tm ice) yield wide-ranging results. Tm ice between -2.2 and -3.3°C were measured in the sample at 257.35 mbmp. In the other sample, at 220.49 mbmp, Tm ice range from -2.2°C to -10.8°C. These temperatures correlate to elevated salinities between 37‰ and 53‰ (mean 46.5‰) and between 37‰ and 148‰ (mean 66.7‰) for those two samples (Fig. 33; Table 9). No normal-marine inclusions were found. Although some of the values appear surprisingly high, they are thought to represent the salinities of the diagenetic fluids, because any re-equilibration or a secondary origin of the inclusions have been carefully excluded.

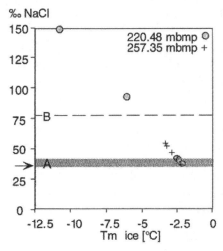

22049 mbmp		25735 mbmp	
°C Tm ice	‰ salinity	°C Tm ice	‰ salinity
-2.8	47	-6.1	93
-3.2	53	-10.8	148
-3.3	54	-2.2	37
-2.45	41	-2.4	40
-2.8	47	-2.5	42
-2.2	37	-2.4	40

Table 9. Tm ice of fluid inclusions examined in lowstand deposits of CLINO, and NaCl equivalent salinities calculated from these temperatures.

Fig. 33. Fluid inclusions in sparry cements of two coarse-grained lowstand sediments and the NaCl-equivalent salinities. Hatched pattern (*A*) marks the range of salinities typical for the present-day Great Bahama Bank (from Traverse and Ginsburg, 1966). Dashed line (*B*): salinity reported from restricted areas west of Andros Island (from Queen, 1978). Arrow marks salinity of normal-marine water.

5.3

Discussion - Diagenesis of Periplatform Carbonates

5.3.1
Diagenetic Environment

5.3.1.1
Stable Isotopic Signatures

In fine-grained carbonates, a distinction between meteoric and marine diagenesis is frequently supported by analysis of stable oxygen isotopes, where meteoric diagenesis is typified by depleted values (James and Choquette, 1984). The $\delta^{18}O$ values of diagenetically altered carbonates are mainly controlled by the temperature at the time of alteration, and by the $\delta^{18}O$ composition of the pore waters. In meteoric fluids, the $\delta^{18}O$ composition is depleted (Allan and Matthews, 1982), and usually is enriched in fluids influenced by evaporation (Anati and Gat, 1989; Ferronsky and Brezgunov, 1989). The $\delta^{13}C$ more often reflects the isotope signal of the precursor sediment. Organically derived carbon in the pore waters, e.g., from soils, however, shifts the $\delta^{13}C$ composition to negative values (Allan and Matthews, 1982).

Melim et al. (1995; in press-b) and Melim (1996) have shown that with respect to isotope composition, the entire core CLINO can be described to consist of two distinct units (Fig 24). The negative $\delta^{18}O$ and $\delta^{13}C$ values present in the upper interval (<135 mbmp) are indicative of diagenesis in meteoric fluids with typical depleted oxygen isotope compositions, and with soil-derived, depleted carbon isotope compositions (Melim et al., 1996, in press-b). The parallel shift of $\delta^{18}O$ and $\delta^{13}C$ towards heavier values between 135 and 153 mbmp is interpreted by these authors to mark the lower limit of meteoric influence during sea-level lowstands. The higher average values in the interval below 153 mbmp indicate diagenesis in marine-derived fluids.

In the present study, the isotopic composition of bulk sediment samples has been determined. The interpretation of bulk samples yields the difficulty that such samples are composed of a mixture of different carbonate minerals (aragonite, dolomite, calcite), that additionally are partly of primary and partly of secondary origin. Individual diagenetic components were difficult to sample, and the subsamples from larger cements and dolomitized components, that were obtained using a dental drill, also could contain multiple carbonate phases.

Oxygen isotopes. The constantly positive oxygen isotopic values (with one exception) for matrix samples obtained in the present study support the interpretation of Melim et al. (1995, in press-b) of diagenesis in marine-derived fluids (-0.07‰ to 3.31‰ $\delta^{18}O$, average 0.72‰ in the Upper Pliocene; 0.63‰ to 1.56‰ $\delta^{18}O$, average 0.97‰ in the Lower Pliocene; Figs. 24, 25, and 26). As noted by Melim et al. (in press-b), the low variability of the isotope values within the interval below 153 mbmp reflects variations in the mineralogic composition of the bulk samples (Fig. 27). The $\delta^{18}O$ of samples with the highest aragonite values (Upper Pliocene samples with up to 92% aragonite) plot around 0.0‰. These values plot close to values from recent bottom sediments from Great Bahama Bank that average around 0.1‰ (Fig. 27G; Shinn et al.; 1989). Usual values for recent aragonite precipitated in normal-marine water at 25°C are expected at -1.0‰ (Grossman and Ku, 1986). The heavier composition observed by Shinn et al. (1989) is considerably influenced by evaporation of the platform top fluids the aragonite muds precipitated from. The enriched values of the aragonitic Upper Pliocene periplatform samples from CLINO could similarly reflect evaporation in the source area of the sediment. Possible hypersaline conditions (elevated saline sea-water) on the platform top are also indicated by the occurrence of dinoflagellate cysts in the sediment that are indicative of such elevated salinity conditions (see Chapter 4).

With decreasing aragonite concentrations (i.e., with increasing diagenetic alteration and increasing low-Mg calcite content), $\delta^{18}O$ values increase towards 1.0‰. This is the opposite to what is expected from recrystallization of aragonite to low-Mg calcite, where a shift towards lighter values by -1‰ takes place (Tarutani et al., 1969; Grossman and Ku, 1986).

The observed shift of $\delta^{18}O$ towards heavier values with higher (mainly diagenetic) low-Mg calcite concentrations is similar to the isotopic change described by Dix and Mullins (1988-b) in sediments from Little Bahama Bank. Dix and Mullins (1988-b) correlate this shift with an isotopic equilibration of the carbonate sediment with deep-marine, cool waters. Generally, the $\delta^{18}O$ becomes enriched by roughly 1‰ at a 4°C cooling (Veizer, 1992). The samples from CLINO are from a considerably shallower depth of deposition than the samples examined by Dix and Mullins (1988-b; 200-300 m opposed to 800-2700 m). The temperature decrease in the present-day thermocline (below ±100 m water depth; Grammer et al., 1993-b), however, may be sufficient to explain the enriched values observed. The high salinities observed in fluid inclusions in sparitic lowstand grainstones that average at 57‰ (Chapter 5.2.6), however, offer an alternative explanation for the enriched values. For salinities between 0‰ and 80‰, a 1‰ salinity increase results in a 0.1‰ enrichment in $\delta^{18}O$ (Anati and Gat, 1989; Ferronsky and Brezgunov, 1989). The possible influence of elevated saline sea-water is discussed below in Chapter 5.3.5.

Carbon isotopes. The $\delta^{13}C$ values obtained within this study scatter between 3.5‰ and 4.5‰. Similar to the oxygen isotopes, the carbon isotopes, especially those for the selected Upper Pliocene interval, plot close to the values of Shinn et al. (1989) that vary from 3.8‰ to 5.0‰ (Fig. 25G). The $\delta^{13}C$ values seem to be less sensitive to diagenetic changes as is indicated by the absence of a clear correlation with the aragonite content. Allan and Matthews (1982) and Joachimski (1994) have shown that in a rock buffered system, the original signal (carbon isotope composition) can be preserved. In the Upper Pliocene, this seems to apply to samples with low permeability (mud- to packstones). In contrast, initially highly permeable, coarse-grained Upper Pliocene lowstand samples exhibit distinctly lower carbon isotope values. Depleted carbon isotope compositions are found in thermocline waters (Grossman and Ku, 1986), therefore no exotic carbon (e.g. from soils) is required for the observed trend towards lower values. The values of the coarse-grained Upper Pliocene lowstand samples overlap with the scatters of the Lower Pliocene samples that are also more strongly altered than the Upper Pliocene highstand deposits (Fig. 25). The lowstand sediments of the selected Upper Pliocene interval seem to have lost their original carbon isotopes signal that would have been retained in a closed rock-buffered system. These lowered carbon values of those coarse-grained samples, together with the higher oxygen values, point to a more open diagenetic system in these initially highly permeable layers. This could also be indicated by the presence of amounts of dolomite exceeding the amount explained by high-Mg calcite unmixing to low-Mg calcite and dolomite. The influence of secular variations cannot be excluded on the basis of the data available. However, on benthic foraminiferal isotope curves of Mix et al. (1995-a, 1995-b) and Shackleton et al. (1995), a significant long-term trend from the Lower to the Upper Pliocene intervals is not observed that could explain for the general shift towards enriched values in the Upper Pliocene.

5.3.1.2

Stable Isotopes and Ultrafacies

When plotting $\delta^{18}O$ *versus* $\delta^{13}C$, the ultrafacies defined in Chapter 5.2.5 separate into well-defined clusters (Fig. 34A). Especially ultrafacies 1, 3, 4, and 5 form clear clusters, whereas the transitional ultrafacies 2 and 6 show intermediate values between ultrafacies 1 and 3, and between 4 and 5, respectively. The ultrafacies show a trend from values that are thought typical for more pristine sediments towards values of more altered sediments. This trend correlates with the increasing cementation as observed under the SEM. This means that the overall trend with progressive diagenetic alteration towards enriched $\delta^{18}O$ and depleted $\delta^{13}C$ compositions is reflected on a small scale when comparing the ultrafacies with the stable isotope compositions. Ultrafacies 1 shows the lowest $\delta^{18}O$ and the highest $\delta^{13}C$ composition. The values shift continuously to enriched oxygen and depleted carbon isotope compositions *via* ultrafacies 2, 3, and 4. Ultrafacies 5, finally, shows the most altered values, while 6 is transitional.

The microfacies defined on the basis of compositional analysis and other light microscopic examinations (Chapter 4), in contrast, show no similarly clear separation in the same plot of $\delta^{18}O$ *versus* $\delta^{13}C$ values (Fig. 34B). Several microfacies (6, 9, 12, and 13), nevertheless, exhibit clusters, suggesting that either specific susceptibilities to diagenesis are inherent in the different microfacies, or that the definition of the microfacies was influenced by features not independent of diagenesis (e.g. diagenetic packstones; see Chapter 5.3.4).

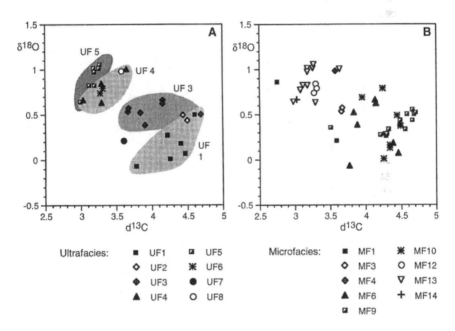

Fig. 34. $\delta^{18}O$ *versus* $\delta^{13}C$ values of samples that have also been classified with respect to ultrafacies and microfacies. **(A)** Note the clear separation of the ultrafacies in clusters showing similar $\delta^{18}O$ and $\delta^{13}C$ values. 30 samples are shown where both, ultrafacies and stable isotopes have been determined. **(B)** Microfacies of the same samples are not that clearly separable. 47 samples are shown where both, microfacies and stable isotopes have been determined. *UF* = ultrafacies; *MF* = microfacies.

5.3.1.3

Significance of Alterations with Respect to the Diagenetic Environment

Specific types of cements are frequently employed in the determination of the diagenetic environment. Especially the meteoric environment in many studies was determined on the basis of specific cements. Among the features that are often thought to be characteristic for the phreatic meteoric environment are blocky spar, dog-tooth spar, and isopachous equant cements composed of low-Mg calcite (Folk and Land, 1975; Flügel, 1982). The clearly marine isotopic signature of CLINO below 153 mbmp has led to the important conclusion of Melim et al. (1995, in press-b) that specific diagenetic features previously interpreted as indicative of phreatic fresh-water diagenesis also occur in the shallow-burial diagenesis in marine-derived pore waters. The present study has not only confirmed these findings, but has also shown that they do not only account for the aforementioned features, but also for microspar that frequently was described as typical for meteoric diagenesis (e.g. Folk, 1974; Steinen, 1982).

The diagenetic alterations of biogenes observed in the present study also closely resemble the patterns described from meteoric environments. In the samples from CLINO, molds of aragonitic components (mainly gastropods and *Halimeda* plates) are common. This type of leaching is well known from meteoric environments (e.g. Dullo, 1983). Neomorphism of aragonite to calcite also was attributed to meteoric diagenesis (Carlson, 1983; Morse and Mackenzie, 1990). In the present study, the neomorphic transformation, that previously was described from CLINO (Melim et al., 1995, in press-b), was looked at closer under the SEM. The skeletal aragonite needles of *Halimeda* fragments are enclosed in cement crystals (Plate 9D), similar to the aragonite needles composing the matrix that are engulfed in microspar. Dissolution prior to cementation probably has played a role as comparably few needles are observed in those neomorphosed *Halimeda* plates. Similar to the neomorphoses, the syntaxial cements are usually thought to be commonly associated with the meteoric diagenetic environment (Meyers, 1978). The syntaxial overgrowths observed in CLINO seem to have formed *in situ* in the marine shallow-burial environment. Although syntaxial overgrowths are more common in the meteoric phreatic environment (Meyers, 1978), they seem to be rather controlled by the availability of the substrate than by the composition of the diagenetic fluid (Melim et al., in press-b).

Thus, the present study supports and supplements the findings of Melim et al. (1995, in press-b) that the marine shallow-burial environment results in similar patterns as the (phreatic) meteoric environment.

In summary, stable isotope examinations have shown that (1) the selected Upper Pliocene interval is less altered than the selected Lower Pliocene interval. The more mature appearance of the selected Lower Pliocene interval at least partly could result from the higher initial permeability. The coarse-grained Upper Pliocene lowstand sediments appear more strongly altered than the fine-grained Upper Pliocene highstand deposits. (2) The six different ultrafacies defined on the basis of SEM observations are reflected in isotopic compositions. (3) The findings of Melim et al. (1995, in press-b) have been confirmed, that diagenesis of the selected intervals of CLINO occurred in marine-derived fluids, and that the significance of diagenetic textural features regarding the diagenetic setting in many cases is limited. In the marine shallow-burial environment similar structures evolve as in the meteoric phreatic environment.

5.3.2
Mineralogic Record

Aragonite and calcite. Droxler et al. (1983) and Reijmer et al. (1988) have shown for Quaternary sediments from the windward side of Great Bahama Bank, that carbonate mineralogy of periplatform deposits potentially can reflect sea-level fluctuations. During highstands, high amounts of aragonitic material are produced on the top of the carbonate platform. This material is shed downslope where the signal is recorded in the periplatform succession (highstand shedding of Droxler and Schlager, 1985). Lowstand deposits lack those high amounts of aragonitic material sourced from the platform top, and mainly consist of bioclasts produced on the lower rim and upper slope, and of open marine biota. Among these grain types, many are of low-Mg calcitic mineralogy. The aragonite-calcite ratio in a vertical section thus can be used to decipher past sea-level changes.

In the sediments examined in the present study, however, the original, potentially sea-level-steered aragonite signal is altered significantly by diagenetic processes. In the selected intervals of CLINO, aragonite is still present, but SEM and light microscopic results imply that the percentages measured with the XRD represent a remnant of the original portion of aragonite. This is also illustrated by the frequency distribution of aragonite in the samples examined. Whereas compositional analysis of the selected Upper Pliocene interval revealed a bimodal distribution of fine-grained matrix with a maximum at 90 % percent matrix (Chapter 4; Fig. 14), aragonite percentages show a unimodal distribution with a peak at 40% (Fig. 35A). This illustrates that the initially aragonitic matrix is diagenetically altered and largely "calcitized". In the Lower Pliocene, generally lower aragonite fractions are present with a maximum at 20% (Fig. 35B) in contrast to the peak of the unimodally distributed fine-grained matrix with a maximum at 60%.

The diagenetic alterations of the components show the same general trend towards mineralogic stabilization as the diagenesis of the matrix. Large amounts of originally aragonitic components are dissolved, or are neomorphosed to low-Mg calcite. These alterations result in a progressive loss of information on the original sediment input. Also, the record of the different biota results in a post-depositional shift in apparent composition, where primary metastable biota are reduced in the fossil record, while initially calcitic biota appear to become relatively enriched (Dullo and Jado, 1984; Dullo, 1990). Not solely initial mineralogy but also grain size influences the rate of diagenetic alteration, small grains being more susceptible to diagenesis (Longman, 1981). The complete absence of coral debris from the fossil record of the successions examined in the present study might at least partly be explained by the extremely high susceptibility of the initially aragonitic scleractinians to diagenetic alterations that is even enforced by small grain sizes (Dullo, 1984). From the difference in behavior of the carbonate mineralogies originates an inherent bias in the fossil record ("diagenetic sieve" of Dullo and Jado, 1984). In the samples from CLINO, this "diagenetic sieve" clearly is observed, although overlain by a real change in the sediment input (Chapter 4).

In vertical section, the diagenetically overprinted mineralogic signal varies between the Upper and the Lower Pliocene successions examined. The present-day mineralogic composition of the Upper Pliocene succession is more variable than that of the Lower Pliocene. This is similar to the disparate sediment input pattern described in Chapter 4. Whereas the compositional signal of the Upper Pliocene sediments exhibits a strong dependence on sea-level fluctuations, the Lower Pliocene shows a rather uniform composition.

Those compositional differences are reflected in the diagenetic signal. In the Upper Pliocene, the mineralogic variations between the predominantly aragonitic highstand sediments and the predominantly calcitic lowstand deposits mirror the differences in primary permeability rather than the original mineralogic signal. The coarse-grained lowstand deposits are generally characterized by high percentages of calcite. This high fraction of low-Mg calcite reflects the presence of sparry cement, large amounts of microspar, and neomorphous components in these diagenetically mature deposits. Pore water flux through the initially highly permeable sediment is thought to have allowed for the dissolution and the transport of calcium carbonate, that precipitated as a calcite cement. The fine-grained highstand deposits, in contrast, appear diagenetically more immature. They exhibit significantly higher aragonite contents, thus reflecting the low initial permeability that hampered rapid pore water exchange. The immature diagenetic state is further proven by the high amounts of aragonite needles as found in many samples under the SEM.

The Lower Pliocene succession, typified by a rather uniform compositional signal, shows generally lower aragonite and higher dolomite percentages. This reflects the overall advanced diagenetic state of the Lower Pliocene interval, that could at least partly result from the overall lower amounts of fine-grained matrix and the resultant higher initial permeability. The mineralogic signal in vertical section of the selected Lower Pliocene interval does not reveal a cyclicity that correlates to the faint sediment input cycles as described in Chapter 4.

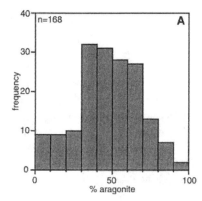 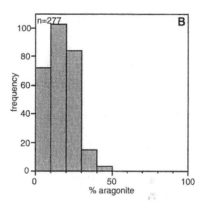

Even though the conditions are not directly comparable, because the CLINO samples are from much shallower water depths, the features observed in the selected intervals from CLINO resemble the observations of Dix and Mullins (1988-b) and Droxler et al. (1988). In their Bahamian periplatform samples from water depth of 800-2700 m, these authors found that at burial depths smaller 10 m, extensive calcitization, aragonite dissolution, high-Mg calcite exsolution, and local dolomitization takes place. Cores from the northern slope of the Little Bahama Bank revealed that during the first 400,000 yr. after deposition, considerable calcitization can occur in a shallow-burial environment (Mullins et al., 1985-b). According to Dix and Mullins (1988-b), this early, very shallow-burial diagenesis is followed by a second stage of shallow-burial diagenesis (<250 mbsf) where slower calcitization and lithification continue. Contrary to Dix and Mullins (1988-b) who observed a parallel loss of aragonite and high-Mg calcite in the top 10 m of the cores off Little Bahama Bank, high-Mg calcite is entirely absent from the intervals of CLINO examined here, while aragonite persists. Periplatform sediments are well known for their high diagenetic potential that is the result of the high portion of aragonite and high-Mg calcite that are metastable under the conditions of deep, cold sea-water (Schlager and James, 1978; James and Choquette, 1983-b). Although the site of deposition for the CLINO sediments was much shallower, above the lysocline, the high amounts of aragonite found in CLINO down to the total depth of the interval examined may appear somehow surprising. In the setting examined here, aragonite dissolution can take place during shallow-burial diagenesis (e.g. Melim et al., 1995). The preservation of metastable minerals is dependent on the relative rates of burial and diagenesis (Friedman, 1965; James and Choquette, 1983-a). Dix and Mullins (1988-b) found that the shallow-burial diagenesis of periplatform sediments occurs rapidly in regions of moderate accumulation rates (e.g. northern slope of the Little Bahama Bank; 20-60 m/Ma). Significantly greater rates result in incomplete diagenesis, because the residence time of the sediment in the active diagenetic environment decreases. Exuma Sound sediments with accumulation rates of 82-210 m/Ma show more gradual changes, and metastable minerals are preserved to greater depths (Dix and Mullins, 1988-b). The accumulation rates calculated for the intervals from CLINO are even higher. They amount to 120 m/Ma for the Lower Pliocene succession, and to around 550 m/Ma for the Upper Pliocene succession (Lidz and McNeill, 1995-a). The immature appearance of large parts of the intervals examined, and the high amounts of aragonite preserved, could result from these high accumulation rates. As Dix and Mullins (1988-b) note, the preservation of metastable minerals after lithification retains a higher diagenetic potential for burial diagenesis that may lead to the (in terms of reservoir evolution) significant secondary vuggy porosity.

Dolomite. When plotting the fractions of aragonite *versus* those of dolomite of the Upper and the Lower Pliocene samples (Fig. 29), it becomes obvious that these two mineralogies tend not to occur together in high amounts. The trend to higher dolomite and lower aragonite fractions from the Upper to the Lower Pliocene interval is a product of progressive diagenesis and could reflect the higher initial permeability of the Lower Pliocene deposits. An exception are the coarse-grained lowstand deposits of the Upper Pliocene that exhibit extremely high dolomite fractions. The initial permeability appears to be a decisive factor for dolomitization.

The different types of dolomite, namely isolated rhombs of micro-dolomite, matrix replacement, and dolomite envelopes, suggest that dolomitization was not a single process but occurred in several episodes. The small isolated dolomite rhombs as seen in the aragonite needle matrix are thought to originate from unmixing of primary high-Mg calcite shed from the shallow-water carbonate factory (see Rush and Chafetz, 1991). This process could account for the low percentages of calcian dolomite as observed in many Upper Pliocene samples. Similarly, Dix and Mullins (1988-b) interpret low amounts (<5%) of small euhedral calcian dolomite rhombs that they found in the zone of early shallow-burial diagenesis (<10 m), as a product of exsolution of high-Mg calcite. Early dolomitization of the sediments recovered in CLINO is supported by Sr-isotopes that yield the same ages as biostratigraphy indicates (Swart et al., in press-a).

The high percentages of dolomite of up to 70% of the total weight of some coarse-grained Upper Pliocene lowstand deposits clearly cannot be explained by diagenetic unmixing of high-Mg calcite. In a sediment originally consisting of 100% high-Mg calcite with 12.5% $MgCO_3$ that, in a closed system, is completely exsolved to low-Mg calcite and dolomite, 25% of dolomite are produced if all Mg^{2+} released were incorporated into the dolomite. In a sediment with aragonite portions as high as seen in the periplatform carbonates of CLINO, not more than a few percent of dolomite are expected to result from unmixing of high-Mg calcite. The dolomite covers around foraminifers (Plate 8 C and D) and other grains observed in the Upper Pliocene lowstand deposits closely resemble specimens described from hypersaline supratidal regions of the modern Bahamas (Lasemi et al., 1988). The dolomite-covered specimens in CLINO, however, clearly are not shed from the platform top. On the basis of textural evidence, for example delicate covers being shared by several grains and the absence of mechanical destruction caused by transport, the dolomite crusts in the samples from CLINO are interpreted to have precipitated *in situ*. Additionally, planktic as well as benthic foraminifers are covered by dolomite, also pointing to an *in situ* origin of the dolomite precipitates.

As a source of the magnesium required to produce such high amounts of dolomite, marine-derived interstitial pore water is possible (Swart and Guzikowski, 1988). Significant flux rates or significant enrichment of Mg^{2+} in the pore water would be required. Melim et al. (in press-b) describe a hardground in the highly dolomitic coarse-grained sediments around 256 mbmp. The formation of a hardground could explain the dolomitization observed. This hardground, however, that according to Melim et al. (in press-b) was subsequently eroded, was not confirmed in the present study. Alternatively, reflux of bank top-derived brines could have provided the Mg^{2+} (see Chapter 5.3.5).

The depositional facies, namely permeability, clearly controls dolomitization. This is seen in the different degrees of alteration in highstand and in lowstand sediments of the Upper Pliocene, and in fine-grained, but component-rich Lower Pliocene deposits. The fine-grained Upper Pliocene highstand deposits contain low amounts of dolomite as expected from (local) unmixing of high-Mg calcite. The coarse-grained, initially highly permeable Upper Pliocene lowstand deposits, in contrast, show varying, high amounts of dolomite that require at least some fluid flux. The somewhat higher initial permeability of the Lower Pliocene sediments could account for the higher dolomite contents than those of the Upper Pliocene highstand deposits. Other factors such as the initial amount of high-Mg calcite and the composition of the pore waters presumably also play an important role in the dolomitization process.

Celestite. Celestite was found mostly, but not exclusively, in cemented layers. The Sr^{2+} required for the formation of celestite ($SrSO_4$) is interpreted to be derived from dissolution of aragonite. Local aragonite dissolution in an aragonitic sediment with 10,000 ppm Sr is sufficient to produce less than 2% celestite if enough SO_4^{2-} is available (Swart and Guzikowski, 1988). Larger amounts of celestite would require major fluid movement through the sediment. Some fluid movement is indicated by the nodular form of the celestite. This nodular form, that is typical for celestite in deep sea deposits (Baker and Bloomer, 1988; Swart and Guzikowski, 1988), however, renders assessment of the amounts of celestite difficult.

The formation of celestite requires SO_4^{2-} and thus its presence is indicative for a marine diagenetic environment. Celestite occurrences have been described from shallow- to deep-burial settings (Baker and Bloomer, 1988; Swart and Guzikowski, 1988). The occurrence of molds in celestite precipitates implies that it precipitated as an early cement. Samples with early celestite cement are frequently uncompacted (ultrafacies 3 and 7) and therefore precipitation of celestite prior to compaction is probable.

Phosphate. Phosphate occurs as fine-grained matrix and as envelopes around components. It is observed under the SEM in only one sample at 509.17 mbmp. This phosphate occurrence implies that the sediment was deposited during times of strongly reduced sedimentary input. This corresponds to the observation that this sample is distinguished by a strongly open marine signature that is interpreted as drowning characteristics (microfacies 14, Chapter 4). Phosphatized horizons are typical for drowning events (e.g. Glenn and Kronen, 1993; Blomeier, 1997). At 27.16 m below the phosphatized sample, a prominent hardground occurs at 536.33 mbmp, that is below the interval examined. The hardground was interpreted as a hiatus covering 3 Myr (Swart et al., in press-a). Above this hardground, phosphatized horizons are numerous (Melim et al., in press-b). This interval coincides with the major backstepping and temporary drowning that resulted in the ramp-type morphology of the Lower Pliocene Great Bahama Bank (Eberli et al., 1997; Kenter et al., in press). The phosphatized layer covered by the sampling for the present study marks the close of the drowning phase that is followed by the catch-up phase of the Lower Pliocene (see Chapter 4).

5.3.3
Lithification Model

Fine-grained limestones have been subject to investigations since the pioneer work of Sorby (1879). Sorby thought the constituents of fine-grained carbonates to originate from the decay of calcareous skeletons into crystallites ("Sorby-principle" of Folk, 1965). The limit of resolution of light microscopes for a long time hampered further investigations of the fine-grained constituents. On the basis of light microscopic examinations, Folk (1959, 1965) proposed a subdivision of fine-grained calcareous constituents into micrite, that is not further resolvable under the light microscope (microcrystalline calcite; <4 µm), and microspar, where single crystals can be distinguished (>4 µm to 30 µm). While the definition of micrite is independent of the origin of the constituents, Folk suggested that microspar forms by recrystallization from a previously lithified micrite (aggrading neomorphism, Fig. 36). Mg-ions, that are released into the pore water during early diagenetic alteration of high-Mg calcite to low-Mg calcite, form an "Mg-cage" around the micrite crystals and thereby inhibit growth of these crystals (Folk, 1959, 1965). When the Mg-ions are removed from the pore water, for instance by fresh water infiltration, the micrite crystals start to grow until they reach microspar size (Fig. 36). Until today, Folks interpretation of microspar formation is widely accepted.

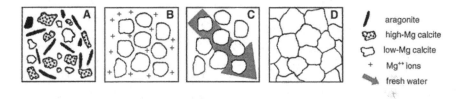

Fig. 36. Microspar development as assumed by Folk (1959, 1965, 1974). (A) Precursor sediment, (B) recrystallization to low-Mg micrite, Mg^{2+}-ions are released to the pore water, (C) fresh water flushes through the sediment and removes the "Mg-cage", (D) aggrading neomorphism to microspar (redrawn after Folk 1974).

A verification of the model of Folk was not feasible as long as the light microscope was the highest resolution tool available to examine the sub-30 µm fraction. The technical development of transmission and scanning electron microscopy (TEM and SEM) in the 1960s permitted examination of textures below the limit of resolution of light microscopes (Fischer et al., 1967; Flügel, 1967; Honjo, 1969; Keupp, 1977, Steinen, 1982). Under the SEM, Lasemi and Sandberg (1984) recognized in Pleistocene aragonite-dominated carbonate muds from the Bahamas and South Florida, that microspar can be formed as a cement by meteoric diagenesis. Calcite crystals with mean diameters between 5 and 15 µm (microspar) are precipitated within the sediment, and small carbonate grains (e.g. aragonite needles) are engulfed in these microspar crystals. This process is completely different from aggrading neomorphism *sensu* Folk (1959).

A study of Munnecke and Samtleben (1996) presented phenomena in Paleozoic micritic limestones from Gotland that resemble the features described by Lasemi and Sandberg (1984). The Silurian limestones from the limestone-marl alternations on Gotland are distinguished by an exceptionally good state of preservation. Diagenetic alterations seem to have ceased shortly after lithification. Thereby many early diagenetic features are preserved that in most pre-Tertiary limestones are lost. The study of Munnecke and Samtleben (1996) revealed that the micritic limestones are dominated by amoeboidal microspar that morphologically closely resembles the microspar described by Lasemi and Sandberg (1984). The microspar crystals observed by Munnecke and Samtleben (1996) are characterized by pitted structures. Such features have been interpreted by Lasemi and Sandberg (1984) to result from dissolution of embedded aragonite needles. Following this reasoning, Munnecke and Samtleben (1996) interpreted the microspar crystals as an early cement that surrounded the sedimentary components and engulfed aragonite needles. On the basis of such observations, they postulated a diagenetic path for the lithification of fine-grained limestones. Lithification by microspar precipitation is thought to have occurred close below the sea-floor, as compactional features are absent from the limestones. The preceding steps of lithification, that are not preserved in the in the Silurian samples from Gotland, were postulated.

SEM examinations of the fine-grained samples from the selected intervals of CLINO revealed striking textural similarities to the Pleistocene samples of Lasemi and Sandberg (1984) and the Silurian samples of Munnecke and Samtleben (1996). Like the Pleistocene and the Silurian samples, the uncompacted samples from CLINO (ultrafacies 3 and 4) are composed of large amounts of microspar. In contrast to the Pleistocene samples, both the Silurian succession and the Pliocene sequences from CLINO were lithified in the marine shallow-burial environment. Therefore, an analogous development of the microspar for the two latter appears likely. In contrast to the Silurian samples, however, in the Pliocene samples metastable constituents (such as aragonite needles) are still preserved. The close textural similarities between the Silurian limestones and the cemented Pliocene samples, and the immatureness of the Pliocene samples indicate, that the latter might represent some of the hypothetical early states in the diagenetic path postulated by Munnecke and Samtleben (1996). The immature diagenetic state preserved in CLINO yields information on the process of lithification that usually is lost in diagenetically more mature limestones, and thereby offers the opportunity to reconstruct the diagenetic processes that led to the microspar textures. The precursor sediment is not preserved in the selected Pliocene interval (Munnecke et al., 1997; Fig. 37) and was hypothesized on the basis of observations of sedimentary constituents in the Pliocene samples. In the following, the diagenetic path is presented, that was reconstructed based on the comparison between the Silurian limestones and the cemented Pliocene samples. For a detailed comparison between the textural features of the Pliocene and the Silurian successions, that was undertaken within the scope of the present study, the reader is referred to Munnecke et al. (1997).

(I) Soft sediment (Fig. 37A). The precursor sediment was composed of an irregular mesh of aragonite needles, less abundant larger biogenic components, and small (partly calcitic) crystallites. The initial mineralogy of the sediment was dominated by aragonite and subordinate high-Mg calcite. The small crystallites at least partly originated from the decay of organically precipitated shells. As Bandel and Hemleben (1975) have shown, biocrystals generally are smaller than 4 µm, and leaching and degradation result in the formation of diagenetically induced micrite ("Sorby-principle" of Folk, 1965).

The composition of the soft precursor sediment was inferred from the Upper Pliocene samples from CLINO that show ultrafacies 1 characteristics (Plate 12A).

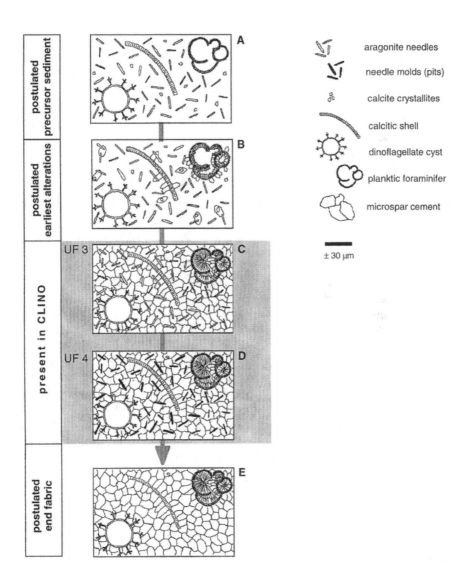

Fig. 37. Cementation of fine-grained, initially metastable periplatform carbonates by microspar precipitation. From these cemented layers, compaction is absent. Most of the Silurian samples correspond to ultrafacies 4 (UF 4). (Based on Munnecke et al., 1997).

The composition of ultrafacies 1 is interpreted to be the closest to the pristine sediment, although compaction and dissolution features clearly indicate diagenetic alterations. Micro-dolomite rhombs in the sedimentary matrix resulted from the alteration of high-Mg calcite precursors to low-Mg calcite and dolomite. The unmixing of initially high-Mg calcitic components seems to have started prior to lithification. In lithified samples, dolomite rhombs, like other components, are enclosed in the cement crystals, indicating that they crystallized prior to cementation. In unlithified as well as in lithified samples, high-Mg calcite components are altered to low-Mg calcite that contains micro-dolomite inclusions (Plate 9E and F).

(II) Lithification (Fig. 37B and C). Small low-Mg calcite crystals nucleated in the sediment. The calcite crystals started to grow on calcitic bioclasts as well as on crystallites presumably derived from the disintegration of calcitic organisms or from epicellular biomineralization (Plate 7D). These cement crystals continued to grow in the crystallographic orientation of their nuclei. When growing into the interparticle pore space, they engulfed tiny components such as aragonite needles (Plate 7D; compare Steinen, 1982; Lasemi and Sandberg, 1984) and fringed larger biogenic components with sharp boundaries (Plate 7E).

Incipient cementation is observed in ultrafacies 2 (Plate 12B). Because of new evidence (compaction indicators), the author of the present study is convinced that, different to what was stated earlier (Munnecke et al., 1997), samples with ultrafacies 2 characteristics do not represent the direct precursor of the cemented ultrafacies, but rather a contemporary, spatial transition between cemented and uncemented samples. Ultrafacies 2 represents an early state of cementation, but compactional features such as deformed dinoflagellate cysts imply that ultrafacies 2 could not evolve into the uncompacted microsparitic ultrafacies 3 and 4. Nevertheless, it is thought that ultrafacies 2 is very similar to the hypothesized preceding state of earliest cementation, and that the precursor of the cemented state looked alike the partially cemented samples, except being uncompacted. The precursor itself, however, is not present in the samples examined.

The microspar continued to grow until it formed an irregular dense mosaic without significant interparticle porosity. As observed by Husseini and Matthews (1972) and Lasemi and Sandberg (1993), the size of the microspar crystals thereby probably was predefined by the amount of calcitic nuclei present in the sediment. Small, sediment-free, intraparticle pores (e.g. in ostracods and foraminifers) were cemented by calcite crystals with smooth crystal surfaces (Plate 9G). The transition within single crystals from cemented sedimentary filling (microspar with pits or needles) to sparry cement with smooth crystal surfaces, present in small geopetally filled cavities, indicates that the microspar and the sparry cement originated from the same process (Plate 7G). Diagenesis of highly porous aragonitic bioclasts such as *Halimeda* plates proceeded similar to the cementation of the needle-dominated matrix. Low-Mg calcite precipitated in the intraskeletal pore space enclosing aragonite needles (Plate 9C and D). The cementation led to the neomorphous appearance of many *Halimeda* fragments. The rare internal acicular aragonite cements, that are exclusively found inside fossil tests and around *Halimeda* utricles, like other aragonite needles, are engulfed in microspar cements.

This state of diagenesis is present in the selected Upper and Lower Pliocene intervals from CLINO (ultrafacies 3; 12C). The absence of compaction in the cemented ultrafacies 3 indicates that the soft, aragonite-dominated, precursor sediment did not experience significant aragonite dissolution prior to the precipitation of microspar, and that microspar precipitation occurred early, before considerable overburden accumulated.

(III) Aragonite dissolution (Fig. 37D).— Lithification (II) was followed by dissolution of aragonite. Aragonite needles engulfed in the microspar crystals were dissolved leaving empty pits (compare Steinen, 1982; Lasemi and Sandberg, 1984). The mechanism of dissolution of aragonite needles engulfed by microspar is not yet fully understood. However, because each aragonite needle must have been in contact with at least one other constituent, a path for diagenetic fluids was created in most cases. Larger aragonitic components were dissolved as well, forming molds. The boundaries between low-Mg calcite-components and (now pitted) microspar crystals remained sharp. This diagenetic state is represented by ultrafacies 4 in the Lower Pliocene of CLINO (Plate 12D). A diagenetic state, that corresponds to ultrafacies 4 as defined in the present study, is the typical pitted microspar of the limestones from the Silurian limestone-marl alternations of Gotland that are devoid of aragonite (Munnecke and Samtleben, 1996; Munnecke et al. 1997).

(IV) Aggrading neomorphism (Fig. 37E).— In a fourth step, slight aggrading neomorphism takes place. This step is not observed in the selected Pliocene intervals from CLINO described here. The following description therefore is based on the phenomena from the Silurian samples described by Munnecke and Samtleben (1996) and Munnecke et al. (1997). Due to the accumulating sedimentary overburden, burial pressure and temperature increased. Larger microspar crystals grew slightly at the expense of smaller calcite crystals, thereby obscuring the boundaries of thin shells and calcareous nannofossils. Nevertheless, the size and shape of the microspar crystals were largely determined during the first step of diagenesis (cementation). The aggrading neomorphism subsequent to cementation only slightly altered size and shape of the microspar crystals. Pitted structures in microspar crystals and molds were filled with calcite cement.

The postulated, late diagenetic end member of this diagenetic development, realized in neither the Pliocene from the Bahamas nor the Silurian from Gotland, is the typical, "well sorted" microspar which has lost its textural evidence (such as e.g. pitted structures) that would allow for the reconstruction of the diagenetic processes.

Based mainly on the textural observations under the SEM, it is concluded that the microspar of this study formed by a process that is fundamentally different from the process of aggrading neomorphism proposed previously (Folk, 1959, 1965, 1974). The microspar crystals in the Pliocene samples from CLINO examined in the present study are not the product of recrystallisation of a previously lithified micrite. Size, shape and texture of the microspar crystals show that they represent the early cement that has lithified the pristine aragonite-dominated carbonate mud. Moreover, the process for microspar formation presented here explains why calcitic bioclasts show sharp boundaries with the microspar matrix (compare Munnecke, 1997; and Munnecke et al., 1997). Aggrading neomorphism (recrystallization) fails to explain this texture. Additionally, recrystallization from low-Mg calcitic micrite to microspar is energetically highly improbable. Once a stable low-Mg calcitic composition is reached, little driving force remains for recrystallization (Veizer, 1977; Steinen, 1978; Sandberg and Hudson, 1983). Similar as in the case of the deep slope carbonates of Little Bahama Bank (Dix and Mullins, 1988-a) it is concluded, that the early onset of alterations observed in the samples from CLINO is a result of the high diagenetic potential of periplatform carbonates.

The textural features observed in the Pliocene samples examined here as well as in the Silurian samples from Gotland that we compared to the Pliocene material (Munnecke et al., 1997), suggest that microspar cement *sensu* Lasemi and Sandberg (1984) not only occurs in meteoric diagenetic environments, but also can be formed during early marine shallow-burial diagenesis without the influence of meteoric fluids. Thus, as noted in Chapter 5.3.1, the environmental significance of microspar, that usually is interpreted as a product of meteoric diagenesis (e.g. Folk; 1959, 1965, 1974), similar to the significance of other cement types and to many diagenetic features of various fossils, is limited. This is in accordance with the growing awareness that many ancient marine carbonates never experienced fresh water diagenesis (Bathurst, 1993).

5.3.4
Cemented versus Uncemented Layers

"The central difficulty is tantalisingly familiar - how to cement a carbonate mud while it is still largely uncompacted. (...) Where was the source of such an enormous quantity of CaCO₃ (...)?" (Bathurst, 1970). Since Bathurst formulated this question, the source of the carbonate cement lithifying the sediment remained one of the crucial problems in carbonate sedimentology (see, e.g., Shinn et al., 1977; Steinen, 1982; Morse and Mackenzie, 1993). When considering the lithification process discussed above, this very question arises. The frequent occurrence of uncompacted limestones throughout Earth's history requires early diagenetic import of carbonate cement occluding the primary pores and stabilizing

a rigid framework. Similarly, the absence of compaction from the cemented limestones of the Pliocene successions from CLINO (ultrafacies 3 and 4) requires an external source of carbonate cement. It also implies that cementation occurred before the pressure of overlying sediment was sufficient to cause a mechanical volume reduction of the sediment. Thus, the assumption of an external source of carbonate cement also accounts for the cemented layers examined here. The volumetric increase of 8.0-8.7%, that accompanies the transformation of aragonite to calcite (Schmidt, 1965; Pingitore, 1970), cannot supply sufficient amounts of carbonate cement to completely fill up the primary pore space. Circulating sea-water as a source for the carbonate cement is unlikely because of the huge volume of water required to provide the dissolved calcium carbonate (Enos and Sawatsky, 1981). Based on calculations of reaction kinetics, Morse and Mackenzie (1993) assume a close spatial relationship between dissolution and reprecipitation of calcium carbonate. Therefore, the source of the cement, especially in the fine-grained sediments with a comparably low initial permeability, should be located within the sedimentary succession, and a local redistribution of the carbonate is expected. Similar reasoning has been employed to explain the cementation of the limestones in limestone-marl alterations (e.g. Eder, 1982; Walther, 1982; Munnecke and Samtleben, 1996, Munnecke, 1997).

If the source of the cement was located close to the site of precipitation (ultrafacies 3 and 4), the interlayering, compacted, uncemented layers (ultrafacies 1 and 5) could be considered as possible sources of the calcium carbonate cement. Aragonite could have been dissolved from the now compacted sediment, subsequently transported by diffusion to adjacent layers, and finally reprecipitated as low-Mg calcite (mainly microspar) in the now cemented layers. Dissolution of aragonite elevated the alkalinity of the diagenetic fluids thus initiating precipitation of low-Mg calcite (Dix and Mullins, 1992). The low contents of Mg^{2+} from the dissolved aragonite shifted the chemistry of the pore water towards lower Mg/Ca-ratios so that low-Mg calcite (and not high-Mg calcite) could be precipitated.

5.3.4.1

Compaction Assessment in Fine-Grained Carbonates

For the reconstruction of the diagenetic path, and to localize the source of the calcium carbonate cement, differentiation of compaction and early cementation is crucial. Laboratory experiments on mechanical compaction with fine-grained sediments have shown that carbonate muds exhibit a behavior similar to siliciclastic muds (Terzaghi, 1940; Shinn et al., 1977). However, as mentioned before, many ancient shallow-water limestones lack indications of compaction, thereby pointing to an early cementation (Weller, 1959; Pray, 1960; Shinn et al., 1977). The relative timing of cementation and accumulation of overburden in carbonate sediments decisively influences the amount of volume loss by mechanical compaction.

The assessment of pre-cementation mechanical compaction in fine-grained carbonates yields difficulties. Whereas a number of indicators are frequently used to assess the mechanical compaction of coarse-grained carbonates (breakage of components, deformation of trace fossils, fitted fabrics, orientation of elongated allochems, crushing of emptied micrite envelopes; Meyers, 1980; Bathurst, 1986; Ricken, 1986), in fine-grained carbonates, the assessment of mechanical compaction is more difficult. Experiments have shown that shell breakage, dissolution at grain contacts, and deformation of ooids in micritic limestones decrease as the relative amount of fine-grained matrix increases. Therefore, in mud-supported carbonates, the absence of deformation and breakage of allochems cannot be regarded as indicator for the absence of mechanical compaction (Fruth et al., 1966; Bhattacharyya and Friedman, 1979; Shinn and Robbin, 1983). Even in grain-supported micritic limestones (packstones), lack of both, breakage and dissolution at grain contacts does not necessarily indicate absence of mechanical volume reduction. Mechanical compaction can result in a decrease of the amount of matrix relative to the amount of components. Therefore, a packstone without signs of deformation can be the product of a highly compacted wackestone (Bathurst, 1986). In addition, the rotation of elongated allochems is an uncertain compaction indicator, because it is difficult to distinguish from sedimentary alignment. Similar problems arise when using peloids as compaction indicators since their ellipsoidal shape can be primary as well as compactional. In mudstones and wackestones, the most obvious sign of compaction is the deformation of burrows (Ricken, 1986). But even this most widely used method to determine compaction in micritic limestones yields problems. Selective cementation of bioturbation tubes can lead to an underestimation of compaction. This differential compaction of burrows and the surrounding matrix has to be corrected for (Ricken, 1987). Lasemi et al. (1990) introduced the preservation of microfenestrae as a scanning electron microscopic approach to estimate the mechanical compaction of fine-grained carbonates. Microfenestrae are voids of about 10 μm in diameter commonly observed in Holocene lime muds mainly from tidal flats and lagoons. Cemented microfenestrae found in micritic limestones are regarded as indicators for pre-compactional lithification. Although the origin of recent microfenestrae from gas bubbles and burrows is widely accepted, the interpretation of their fossil counterparts sometimes is problematic. Similar textures can also have originated from dissolved and refilled, initially aragonitic, shell debris. Furthermore, microfenestrae are variable in size and shape, rendering it difficult to assess their deformation. Pre-cementation compaction results in the disappearance of the microfenestrae. Their absence in a limestone thereby suggests mechanical compaction of the sediment (Lasemi et al., 1990). This, however, holds only if microfenestrae had been present in the soft sediment. For periplatform sediments, where they are rare, their application is limited.

In the course of the present study, a complementary tool for the assessment of compaction of fine-grained limestones was proposed (Westphal and Munnecke, 1997). As indicated by a number of different compaction indicators (e.g. deformation of pellets and burrows, breakage of foraminiferal tests, diagenetic lamination), the samples from CLINO show varying states of compaction. Undeformed peloids and bioturbation burrows indicate that the microsparitic ultrafacies 3 and 4 were not subjected to significant compaction prior to lithification. Cementation of this sediment has occurred early, i.e. before the sediment overburden was thick enough to cause compaction. In these layers, dinoflagellate cysts are spherically preserved to slightly deformed (Plate 10A and B).

Samples with ultrafacies 1 and 5 characteristics are discernible by strongly deformed pellets and burrows. Under the light microscope, such samples frequently appear laminated due to compaction, and microspar crystals are rarely seen under the SEM. In these compacted layers, dinoflagellate cysts are strongly deformed (Plate 10C and D) and usually aligned to the (diagenetic) lamination. In the partially cemented ultrafacies 2 and 6, varying states of preservation are observed.

The correlation of the preservation of dinoflagellate cysts with other compaction indicators (peloids, diagenetic lamination, breakage of planktic foraminifers) allows the application of dinoflagellate cysts to assess compaction even where other indicators are absent. The deformation of thin-walled organic microfossils *in situ* potentially could develop into a widely applicable new tool for the assessment of mechanical compaction in fine-grained carbonates. Compared to other methods, the use of organic-walled microfossils as compaction indicators yields the following advantages: (1) The original, uncompacted shape of the microfossils is known. Sometimes perfectly preserved spherical specimens occur in the same sample as slightly deformed ones. If a rather uniform compaction within one sample was effective, it must be assumed that the latter were already slightly deformed when deposited. Therefore, those individuals preserved best are the most reliable indicators for compaction. (2) In mud-supported carbonates, the absence of deformation and breakage of calcitic allochems is not indicative for the absence of mechanical compaction (Shinn and Robbin, 1983). Organic microfossils appear more sensitive to compaction. Their compactional behavior is, similar to that of burrows, not dependent on the mud/allochem-ratio. (3) Unlike microfenestrae, palynomorphs are present in most marine environments, and they do not disappear and are still recognizable when compacted. Thereby, they are straight-forward indicators of mechanical compaction. (4) Furthermore, palynomorphs, although only accessory constituents, are rather easily spotted in SEM by their lighter appearance in moderate magnifications (150x). The investigation of statistically significant numbers of individuals, however, is very time-consuming. Therefore, it is a qualitative, rather than a quantitative, method to differentiate compaction *versus* early cementation.

5.3.4.2
Source of the Carbonate Cement

The compaction assessment supplemented by examining the preservation of dinoflagellate cysts supported that clearly different states of compaction are present in the intervals examined (Fig. 38). Seemingly uncompacted, partly compacted and strongly compacted samples occur in an irregular order. The potential calcium carbonate source rocks (ultrafacies 1 and 5) are distinguished by compactional features. In SEM, the "groundmass" of these two ultrafacies showed up as devoid of cement crystals (microspar). (The two partially cemented microfacies 2 and 6 will be addressed later in this chapter.)

In the Upper Pliocene, the compacted, possible source layers (ultrafacies 1) still contain high amounts of aragonite. Nevertheless, the affected appearance of the aragonite needles as seen in unetched broken surface samples (Plate 4H) implies that dissolution has taken place.

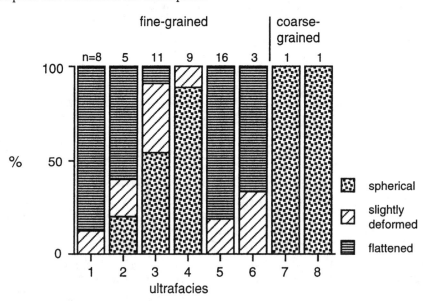

Fig. 38. Deformation of dinoflagellate cysts in the different ultrafacies. Ce-mented fine-grained ultra-facies 3, 4 and coarse-grained ultrafacies 7 and 8 are characterized by a predominance of spheri-cally preserved dinofla-gellate cysts, whereas the uncemented and partly cemented ultrafacies 1, 2, 5, and 6 are typified by deformed to flat speci-mens. n=numbers of samp-les where the preser-vation of dinoflagellate cysts was evaluated.

The process of aragonite dissolution, however, obviously is not completed yet. Some local redistribution of calcium carbonate could be represented by the rare internal cements in foraminiferal tests that possibly formed a micro-environment favoring precipitation. Most foraminiferal tests, however, are devoid of internal cementation (Plate 11H). In the compacted ultrafacies 5 of the Lower Pliocene of CLINO, aragonite fractions are lower than in ultrafacies 1. Further leaching in the compacted layers could have led to a considerable loss of aragonite. Ultrafacies 5 samples are strongly compacted as indicated by the laminated appearance in thin section and by the presence of flattened dinoflagellate cysts as seen under the SEM. The occurrence of pores that resemble the "micro-vugs" of Dix and Mullins (1988-a) and Moshier (1989) suggests that dissolution of components took place. Ultrafacies 5 consists mainly of calcite grains with a smaller average grain sizes (6.5 μm) than the cemented samples (around 7.5 μm). This implies that ultrafacies 5 could at least partly be composed of the calcitic residue of the presumed source of the carbonate cement, that was impoverished in aragonite. The crystallites are thought to be at least partly derived from the decomposition of calcitic shells. The larger crystals observed in ultrafacies 5 (rarely exceeding 14 μm) could have grown by cementation in a later diagenetic stage. The shape of these crystals (roughly euhedral to anhedral) implies that they are inorganic in origin. The partially cemented samples of ultrafacies 2 and 6 could represent spatial (vertical) transitions between cemented and uncemented layers. In the immature successions from CLINO, the boundaries between cemented and uncemented layers could still be transitional, while they could become sharper in further diagenesis. Layers with transitional states of compaction are known from numerous limestone-marl alternations, where they are located between the individual limestone and marl beds (Walther, 1982).

The coarse-grained deposits with their high initial porosities and permeabilities are generally uncompacted, and thus are unequivocal cement importers. In highly permeable sediments, the question of the source of the carbonate cement is more difficult to address, because longer transport distances are possible. The fact that they appear more mature than the adjacent fine-grained highstand sediments underlines the primary differences. Obviously, the originally high permeability of the coarse-grained layers resulted in a high early post-depositional fluid flow, and thus in an accelerated diagenesis. Halos of alteration to low-Mg calcite around grainstone beds in CLINO that are described by Melim et al. (in press-b) emphasizes the influence of high permeability on cementation. Therefore, the lowstand deposits are not readily included in the picture drawn for the fine-grained deposits.

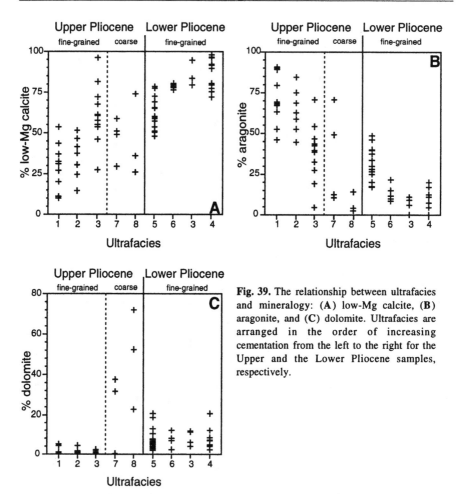

Fig. 39. The relationship between ultrafacies and mineralogy: (A) low-Mg calcite, (B) aragonite, and (C) dolomite. Ultrafacies are arranged in the order of increasing cementation from the left to the right for the Upper and the Lower Pliocene samples, respectively.

The diagenetic differentiation is reflected in small-scale variations in carbonate mineralogy in the order of decimeters to meters, that overlay the larger-scale trends discussed in 5.3.2.1. These small scale variations are found to correlate to the ultrafacies (Fig. 39). Increasing cementation coincides with a decreasing aragonite fraction. Uncemented, compacted, samples (ultrafacies 1 and 5) are characterized by higher aragonite contents than cemented samples (ultrafacies 3 and 4). Ultrafacies 2 and 6 show intermediate values. The tightly cemented, coarse-grained, lowstand samples (ultrafacies 7 and 8) show a wide range of aragonite contents depending on the state of preservation of the initially aragonitic components present. The mineralogic signature reflects the import of low-Mg calcitic cement infilling the initial pore space. Cementation by low-Mg calcite led to "dilution" (relative impoverishment) of aragonite by the import of calcite cement. The dissolution of aragonite in the compacted layers (ultrafacies 1 and 5) is incomplete, leading to a maximum of aragonite in the compacted layers not intuitively expected.

Besides the generally higher amounts of dolomite in the Lower Pliocene, no correlation exists between dolomite content and the different ultrafacies that characterize the fine-grained samples (Fig. 39). This indicates that in the fine-grained samples, the dolomite fraction is not merely a function of passive enrichment of early diagenetic dolomite by compaction. The lack of correlation rather suggests that dolomitization occurred in several episodes. The coarse-grained ultrafacies 7 and 8, in contrast, show distinctly higher amounts of dolomite, indicating that high pore water flux enabled the import of considerable amounts of magnesium.

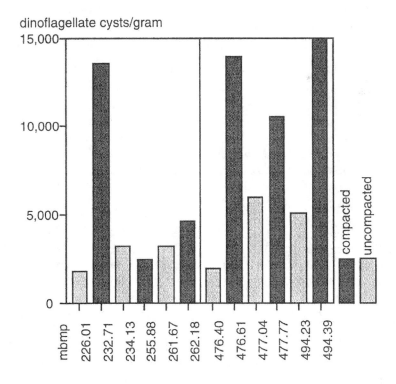

Fig. 40: Quantity of dinoflagellate cysts in compacted and uncompacted fine-grained Upper and Lower Pliocene samples (see table 10). Note the distinctly higher numbers in most compacted samples, pointing to a passive enrichment of insoluble residue.

[mbmp]	compacted?	cysts/g
226.01	no	1,796
232.71	yes	13,511
234.12	no	3,187
255.88	yes	2,446
261.67	no	3,208
262.18	yes	4,576
476.4	no	1,933
476.61	yes	13,897
477.04	no	5,945
477.77	yes	10,571
494.23	no	5,096
494.39	yes	14,977

Table 10: Abundance of dinofla-gellate cysts per gram sediment. Note the on the average higher numbers of cysts in the compacted layers.

The extent of dissolution of aragonite and reprecipitation of calcite is difficult to assess, because terrestrial insolubles such as quartz or clay minerals, that could serve as a reference for quantification, are present in insufficient amounts. Palynologic examinations revealed that palynomorph associations can be employed as an insoluble reference. The species associations present in the different samples do not show any systematic difference between cemented and compacted layers, so that the palynomorphs are regarded as potentially uniform input. Statistically sufficient numbers of individuals in order to evaluate a relative enrichment were gained using standard palynological methods where the calcium carbonate is dissolved. The abundances of palynomorphs in compacted layers mostly exceed those of the adjacent, uncompacted layers (Table 10, Fig. 40). Compacted layers contain in the order of 10,000 dinoflagellate cysts per gram sediment, whereas in cemented layers dinoflagellate cysts show around 3,500 specimens per gram sediment. In the Lower Pliocene succession examined, the compacted layers constantly yield around three times the amount of cysts than the cemented layers (Fig. 40; Table 10). In the Upper Pliocene, the picture is less striking. The reason for the deviations might not only be the more immature state of diagenesis, but also the climatic conditions in the Upper Pliocene that were less constant than in the Lower Pliocene. This passive enrichment of organic microfossils in the compacted layers supports the assumption that aragonite was dissolved from these layers.

The same trend as seen in the quantitative distribution of palynomorphs is observed in the fraction of organic carbon present in the samples. Organic carbon portions are positively correlated with the abundance of dinoflagellate cysts (Fig. 41), and (within each of the two data sets) also with the aragonite fractions (Fig. 30). (The reason for the offset between the measured values from Upper and Lower Pliocene seen in figure 30 is not clear; primary differences in the input influenced by the distance to the platform margin could have played a role.) The passive enrichment of organic carbon in marls was reported from limestone-marl alternations of various ages (Ricken, 1993). Similarly, in the successions from CLINO, the compacted layers could be passively enriched in organic carbon, whereas the organic carbon as well as the aragonite are diluted in the limestones by import of carbonate cement.

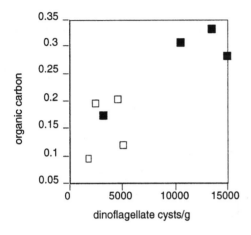

Fig. 41: Positive correlation of dinoflagellate cysts and organic carbon in the samples from Table 10 where data were available, exhibiting different degrees of compaction. Black boxes = compacted; empty boxes = uncompacted. The compacted sample plotting at low dinoflagellate cyst quantities is from the Upper Pliocene.

The hypothesis, that the compacted ultrafacies 1 and 5 are the source of the carbonate cement located within the succession, leads to a diagenetic model that complements the model of lithification by microspar precipitation (Fig. 37). In this model, both possible paths, sediment importers and sediment exporters, are considered (Fig. 42). The precursor sediment is altered in two ways. Progressive dissolution takes place in layers that are being compacted, while contemporaneous cementation occurs in adjacent layers that remain uncompacted. The postulated end fabric of the cementation path is the "well sorted" microsparitic limestone (Chapter 5.3.3). The postulated end fabric of the dissolution path is a highly compacted limestone devoid of aragonite, with possibly smaller grain sizes than its cemented counterpart.

The diagenetic development of cemented and uncemented layers follows two distinct paths that are differentiated early in diagenesis. The differential compaction underlines that the different diagenetic styles (cemented and uncemented) are the result of diagenetic differentiation. In further diagenesis, with increasing overburden the cemented layers will experience no mechanical but only chemical compaction, possibly with the formation of stylolites. The intercalated, uncemented layers will most probably undergo further mechanical, and later chemical, compaction. Thus, the diagenetic paths of compacted and uncompacted layers will continue to diverge.

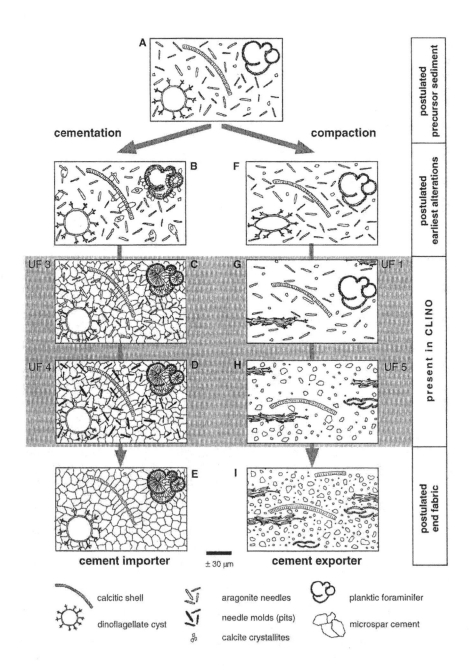

cementation

compaction

postulated precursor sediment

postulated earliest alterations

present in CLINO

postulated end fabric

cement importer   ± 30 µm   cement exporter

	calcitic shell		aragonite needles		planktic foraminifer
	dinoflagellate cyst		needle molds (pits)		microspar cement
			calcite crystallites		

Fig. 42. For caption see opposite page.

Fig. 42. Diagenetic development of fine-grained, initially metastable carbonates in the marine shallow-burial environment. Left column: lithification path of the uncompacted cement importers; from figure 37. Right column: dissolution and compaction of the assumed source layers of the carbonate cement. Decision on the differentiation in the cementation path or the compaction path occurs early in diagenesis.

Since pressure and temperature conditions in the shallow-burial environment are insufficient at the burial depth in question, processes of redistribution of calcium carbonate and mineral stabilization must be triggered by changes in pore water chemistry. In the shallow-burial environment, strong geochemical gradients can result from the microbial decomposition of organic matter (Canfield and Raiswell, 1991). Munnecke and Samtleben (1996) propose that the process of redistribution of calcium carbonate could be triggered by microbial deterioration of organic matter. Such chemical gradients have been observed to cause carbonate solution close to the sea-floor (e.g. Chilingar et al., 1967). The relatively low contents of organic carbon in CLINO could offer an alternative explanation for the incomplete diagenetic appearance of the sediments to the assumption of Dix and Mullins (1988-a, 1988-b) who proposed sedimentation rates as a steering factor. This, however, is difficult to prove as the hydrogeochemical conditions present in a sequence today need not have a close relationship with the fluids responsible for early diagenetic processes.

5.3.4.3
Causes for Diagenetic Differentiation

The causes for rhythmic alternations of cemented and uncemented layers have long been, and still are, controversial. Such alternations have been described from Bahamian periplatform carbonate successions (Schlanger and Douglas, 1974; Dix and Mullins, 1988-b), but are better known from many studies on limestone-marl alternations of various ages and from many locations (for an overview see Einsele and Ricken, 1991). As steering factor of these alternations, sediment input variations have been proposed that are enforced by diagenesis, or opposed to this interpretation, an entirely diagenetic origin was postulated (see, e.g., Bathurst, 1971; Milliman, 1974; Einsele, 1982; Ricken and Eder, 1991). Especially for pure carbonates, the mechanisms of diagenetic differentiation are still far from being thoroughly understood. Schlanger and Douglas (1974) thought that the separation in cemented and uncemented layers in periplatform carbonates could be caused by differences in the diagenetic potential (sediment input cycles). They also considered fluctuations of the carbonate compensation depth and varying sedimentation rates as possible causes. Dix and Mullins (1988-b) interpreted differential diagenesis as an effect of either changes in the relative accumulation rate *versus* diagenetic alteration rate, or similarly to Schlanger and Douglas (1974) as differences in the initial diagenetic potential (aragonite, high-Mg calcite, and/or organic matter contents).

To assess whether a primary signal (sediment input) steered the differentiation in the intervals from CLINO, the correlation between ultrafacies (defined on the basis of diagenetic alterations) and microfacies (defined on the basis of sediment input patterns, Chapter 4) was examined. The selected Pliocene successions exhibit eight different ultrafacies describing the diagenetic appearances, six of which occur in fine-grained sediments. The correlation between the 14 microfacies and the ultrafacies is rather poor (Fig. 43). This indicates, that, although sediment input

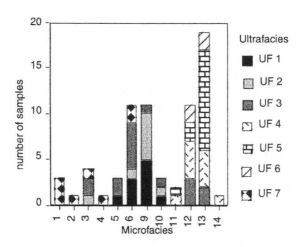

Fig. 43. Distribution of microfacies defined in Chapter 4 that mainly describe compositional signatures, relative to the ultrafacies that describe diagenetic states.

variations influence diagenesis (as seen in the differences between coarse-grained and fine-grained deposits), different ultrafacies could have developed from the same microfacies. Sediment input patterns as the sole cause for the differentiation in cemented and uncemented layers, therefore, is not supported by the data. The absence of systematic trends in the spectrum of palynomorph species in cemented and compacted layers also indicates that a sedimentary signal might not be the steering factor. However, the influence of other environmental signals cannot be excluded on the basis of these observations.

Recent publications (e.g. Munnecke and Samtleben, 1996; Bausch, 1997; Munnecke, 1997) have suggested that the classical definition of limestones and marls in rhythmic alternations (see Correns, 1949), that is based on carbonate contents, is inadequate for many alterations that otherwise satisfy the criteria of limestone-marl alternations. Bausch (1997) has shown that there are rhythmic alternations exhibiting a wide range of carbonate contents, with an extreme "end-member" represented by lithographic limestones that are virtually free of terrestrial material. Munnecke and Samtleben (1996) propose that in limestone-marl alternations the carbonate contents could not be used as characteristics for the definition of limestones and marls, but the fact that limestones are uncompacted whereas marls are compacted. According to these authors, the carbonate contents depend on the relative abundances of calcitic, aragonitic and insoluble contents in the precursor sediment. Precursor sediments that virtually lack insolubles would lead to alternations that, like the lithographic limestones of Bausch (1997), are entirely calcareous, but correspond to limestones and marls. The Pliocene succession from CLINO could possibly represent an immature state of such a rhythmic alternation.

5.3.5
Origin of Diagenetic Fluids

One aspect of the diagenesis of carbonate platforms that is difficult to address is the chemistry of the pore fluids that led to the diagenetic alterations found in the sediments. To obtain information on the pore fluids and fluid flow systems taking part in the diagenesis of a carbonate platform, different approaches are possible. (1) *In situ* measurements of the present day fluid flow in natural systems within carbonate platforms (e.g. karst tubes) allow for the establishment of models that, by analogy, could be applied for the diagenetic features observed in older carbonates (Whitaker and Smart, 1990; Whitaker et al, 1994, 1997). (2) Present day pore waters from carbonate rocks can be obtained by squeezing of samples and by pumping pore waters from boreholes (Swart et al., in press-b). These first two approaches yield the inherent difficulty that pore fluids associated with a carbonate rock today not necessarily need to have a relationship with the diagenetic features observed in the host rock. (3) A third approach that avoids this problem is the examination of fluid inclusions. Fluid inclusions (if they are primary and not re-equilibrated) represent the diagenetic fluids that led to the precipitation of the cement they are enclosed in (Goldstein and Reynolds, 1994). Examination of fluid inclusions has been demonstrated to be a valuable tool for characterizing the diagenetic environment that led to the alteration of carbonates (vadose *versus* phreatic and freshwater *versus* normal saline or elevated saline sea-waters; e.g. Goldstein, 1988, 1990; Goldstein and Reynolds, 1994). This method has been successful, for example, in proving the elevated salinity of diagenetic fluids that led to dolomitization in the deep subsurface of Enewetak atoll (Goldstein and Reynolds, 1994).

In the present study, the third approach is employed to gain information on the diagenetic fluids involved in the alteration of the Pliocene samples from CLINO. This method, however, can only be applied where large enough sparry cements are present. In the case of the selected intervals from CLINO the examination of fluid inclusions is restricted to the Upper Pliocene lowstand deposits. Fluid inclusions in the sparry cements of the lowstand deposits reveal that hypersaline diagenetic fluids (elevated saline sea-waters) have contributed to the early diagenesis of the lowstand deposits. The salinities calculated from the fluid inclusions vary between 37‰ and 148‰ NaCl equivalent (Table 9, Fig. 33). This indicates that at least some growth zones of the cement (but not necessarily all of them) precipitated from a fluid that was more saline than sea-water.

Several mechanisms are capable of generating and driving a saline ground-water flow that could be involved in diagenetic alterations of carbonate platforms. Among those are differences in the hydraulic head caused by tides, local wave conditions, or currents (e.g. Carballo et al. 1987; Atkinson et al., 1981), differences in the temperature (geothermal heat; Kohout et al., 1977), and finally differences in the density (reflux; e.g. Adams and Rhodes, 1960; Simms, 1984). In the system examined here, two models appear possible as dominant mechanisms that could have occurred combined with other, subordinate mechanisms: (1) reflux of elevated salinity waters, or (2) uplift of underlying brines by Kohout convection *sensu* Simms (1984).

(1) Brines could have been generated by evaporation on the platform top and subsequently refluxed through the carbonate platform. The hypothesis of seepage reflux was formulated by King (1947) and Adam and Rhodes (1960). It states that normal-marine waters, that are concentrated at the surface of a platform by evaporation, could move downwards within the platform and displace less dense pore fluids at depth. The main driving mechanism would be the development of hypersaline lagoonal waters. Simms (1984) has shown that already just slightly elevated salinity can be sufficient to initiate reflux. This was proven by Whitaker and Smart (1990) who demonstrated that reflux is part of the present-day active circulation system in Great Bahama Bank. Restricted circulation is common on rimmed platforms, but also occurs on platforms that lack a rim if they are areally extensive and shallow (Simms, 1984). Salinities of 37‰ to 42‰ are observed on the present-day platform top of Great Bahama Bank (Traverse and Ginsburg, 1966; Fig. 44), even though rainfall averages about 114 cm/year (Dravis, 1977). Off the westcoast of Andros Island, salinities locally exceed 45‰ (Bourrouilh-le-Jan, 1980), and salinities even exceeding 80‰ have been reported from restricted regions close to the shore of Andros Island (Queen, 1978). To the west of Andros Island, the residence time of marine waters on the platform top is about one year, thus resulting in progressive concentration by evaporation (Broecker and Takahashi, 1966).

The highly concentrated fluids found in some fluid inclusions in the Upper Pliocene lowstand deposits could have been generated during periods of aridity and poor circulation while sea-water covered the platform top. The most effective restriction on the platform top could have occurred during times of initial flooding of the platform, in a late highstand phase, or during the initial fall of sea-level. The elevated salinity waters found on the present-day platform top (42‰) are denser by about 5 g per liter than the slightly elevated saline waters of the adjacent Straits of Florida (37‰), even though the first have higher water temperatures (calculated after Dietrich et al., 1975). The denser saline waters could have migrated downwards *via* permeable layers. The coarser-grained lowstand deposits could have served as aquifers for such fluids, which may eventually have discharged on the slope. On theoretical grounds, Simms (1984) calculated that depths of 300 m below the bank top could potentially be flushed by large-scale reflux of saline fluids from Great Bahama Bank. On the basis of *in situ* measurements of the composition of waters in the karst system of Great Bahama Bank, this value was corrected to 168 m for the present-day system (Whitaker et al., 1994). These values compare roughly to the depth of deposition and early burial estimated for the selected Pliocene periplatform intervals from CLINO.

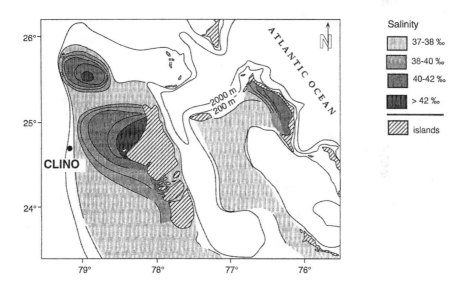

Numerical models of Kaufman (1994) suggest that the distances and rates of density-driven reflux of fluids and solutes are a function of the degree of evaporation. Whitaker et al. (1994) demonstrated that the present-day circulation within Great Bahama Bank is dominated by two types of waters, elevated salinity waters from the bank top, and normal salinity water from depth of the adjacent ocean. The variations in the salinity measured in the fluid inclusions from CLINO could be explained by variations in the mixing of similarly different water masses.

The abundant dolomite found in some coarse-grained lowstand deposits is consistent with the reflux model, because an evaporated fluid would provide a source of magnesium ions. Also, the presence of typical halophil dinoflagellate cysts (especially *Polysphaeridium zoharyi*) in the periplatform sediments indicates that elevated salinities prevailed on the Pliocene platform top. The positive oxygen values in some calcite cements (around +1.0 $\delta^{18}O$), however, would be in accordance with both, cool, normal-marine diagenetic fluids and refluxed, elevated salinity fluids, as both could exhibit similar values.

Low-Mg calcite cements are a diagenetic phase not normally associated with reflux of evaporated sea-water. However, it could be hypothesized that the brines could have dolomitized carbonate on their way down. Mg^{2+} could have been removed from the fluid, and Ca^{2+} could be have been added by this reaction, eventually driving the fluid towards supersaturation with respect to low-Mg calcite.

(2) The second possible source of the high-salinity fluids found in the fluid inclusions are underlying deep-seated evaporites or brines. Pore water measurements of ODP Leg 166 samples revealed that salinities in the lower slope to toe-of-slope sediments increase downhole to considerable depths in the order of 1000 m below sea-level (several 100 m below sea-floor; Eberli et al., 1996; Shipboard Scientific Party, 1997). Brines from dissolution of possible underlying evaporites or underlying brines of unknown origin could be the source of these saline fluids (Eberli et al., 1996). Such elevated salinity fluids could have been brought up by Kohout convection. Because of the downhole increase in salinity, this model is favored by the Shipboard Scientific Party (1997). Also, reflux usually only extends down to depths of up to 300 m (Simms, 1984; Whitaker et al., 1994), but the brines extend down to considerably greater depths. The problem in the application of the Kohout model, however, is the improbability of an upward flow of such dense, highly saline brines as found in the fluid inclusions measured in the present study. A mixed source origin appears possible with parts of these deep-seated saline fluids being derived from the platform surface (Eberli et al., 1996).

The fluid flow system in Great Bahama Bank is not yet fully understood. A progress in the understanding of the fluid flow system can be expected from further studies that are currently carried out, e.g. by P. Swart, on the ODP Leg 166 data.

If the salinity of the fluid flow, at least in the shallower regions of the slope, is considered to be entirely or partly steered by reflux of elevated salinity waters, then a dependence on sea-level fluctuations is given. Even though the direct influence of sea-level fluctuations is restricted to the platform top, sea-level position controls the generation of concentrated brines by evaporation on the platform top. Such fluids in turn could reflux into the slope sediments and influence their diagenesis.

As was shown, sea-level position also controls the mineralogy and the texture of the slope sediments (Chapter 4) that later affect diagenetic alterations. The compositional differences mainly result in differences in the cementation and the diagenetic matureness of a sediment. Light microscopic and SEM analyses (Chapter 4) have shown that highstand deposits are represented by micritic carbonates with high amounts of aragonite needles, whereas lowstand deposits consist mainly of grainstones with coarser components such as *Halimeda* plates and cortoids. These compositional differences result in differences of the primary porosity and permeability. Highstand deposits, being fine-grained, are characterized by low initial permeability. They are diagenetically immature and contain high amounts of aragonite. Their initially low permeability precluded rapid fluid migration and protected these sediments from fluid-driven diagenetic alteration. These fine-grained deposits appear to be cemented by locally derived calcium carbonate from intercalated, compacted layers (Chapter 5.3.4). The diagenetic fluid could have been derived from normal-marine water. Incomplete dissolution of aragonite in compacted samples leaves highly aragonitic samples (like those of ultrafacies 1) for further diagenesis, such as later leaching that could result in considerable porosity. High strontium contents in pore water samples from CLINO indicate that diagenetic processes are still proceeding (Swart et al., in press-b). Unfortunately, the intervals selected for the present study are not covered by water samples of the study of Swart et al. (in press-b), therefore no direct comparison was possible.

Lowstand deposits are characterized by a higher initial porosity and permeability than the fine-grained highstand deposits. This is reflected in the diagenetic maturity. The coarse-grained lowstand deposits of the selected Upper Pliocene interval are usually tightly cemented and contain low amounts of aragonite. Dolomite occurs in strongly varying amounts and can exceed 70%.

As a predictive model, one would expect that in micritic highstand slope deposits, considerable amounts of metastable aragonite survives early diagenesis. These sediments would be capable of forming considerable secondary porosity by later diagenetic alteration.

Typically, the lowstand deposits have higher initial porosities, and could thus exhibit a high potential for at least preservation of some primary porosity. However, the high initial permeabilities could have focused diagenetic fluids during early diagenesis. This high fluid flux could result in rapid diagenetic alteration. Given a model of reflux, primary porosity is probably best preserved in settings that remained humid, never generating high salinity fluids on the banktop. For those banks that had generated such fluids, diagenesis in updip positions of the lowstand deposits are most likely to dolomitize the sediment and potentially may preserve some porosity. However, in downdip positions of lowstand deposits, it is expected that low-Mg calcite cementation by saline fluids takes place, that dolomitized the updip portion of the lowstand slope deposits. These sediments would likely not retain reservoir porosity.

5.4

Conclusions

Examinations of the diagenetic patterns in the Pliocene periplatform carbonates from CLINO led to the following conclusions:

(1) In the selected Upper Pliocene succession clearly the compositional signatures of sea-level fluctuations, namely initial porosity and permeability, influenced the early diagenetic patterns. Fine-grained sea-level highstand deposits have a low primary permeability. They show an immature diagenesis with uncemented, partly cemented and completely cemented layers. Coarse-grained sea-level lowstand deposits, in contrast, have had high initial permeabilities and therefore were subject to greater fluid flux. They are usually tightly cemented and uncompacted, and can contain high amounts of dolomite.

The dependence of diagenesis on primary permeability is a link between diagenesis and sea-level fluctuations, at least for flat-topped carbonate platforms with distinct changes in composition.

(2) The nature of the diagenetic fluids involved in early diagenesis seems to be also influenced by the initial permeability. In fine-grained deposits, a local redistribution of calcium carbonate could have been mediated by marine-derived pore fluids. The initial low permeability of these sediments precluded rapid fluid migration and protected these sediments from fluid-driven alteration.

Coarse-grained deposits with high initial permeabilities were subject to greater fluid flow. Elevated saline fluid inclusions indicate that one type of fluid that may have been important for dolomitization and low-Mg calcite cementation of coarse-grained deposits was an elevated salinity brine. This brine could have formed by evaporation on the bank top during times of restriction associated with initial flooding, late highstand or initial fall in sea-level. The high salinity fluids were dense enough to sink into the platform and migrate through the permeable slope sediments deposited during lowstands. This, among other mechanisms of fluid flow, was responsible for the intense diagenetic alteration of lowstand-deposited slope sediments.

(3) A variety of diagenetic patterns that are usually interpreted as indicative of meteoric diagenesis, also occur in the marine shallow-burial environment. This not only accounts for sparry cements (Melim et al., 1995, in press-b) but also for microspar formation and for diagenetic patterns of various bioclasts.

(4) In the Pliocene periplatform carbonates examined, microspar is an early cement lithifying the aragonite-dominated mud. It is not a recrystallized micrite, the product of aggrading neomorphism *sensu* Folk (1959). Comparison of Pliocene periplatform carbonates from the Bahamas with Silurian limestones from

Gotland reveals that, despite differences in sediment composition and age, an analogous lithification process has occurred. The similarity of diagenetic environments lithifying these carbonates allows for the reconstruction of microspar development from unconsolidated aragonite-dominated mud to consolidated microsparitic limestone in a shallow-burial environment. Neither high pressure nor high temperature are required for the formation of microspar.

The formation of microspar can occur as an early step in limestone diagenesis. Microspar formation and the subsequent dissolution of enclosed aragonite constituents (resulting in pitted microspar crystals) is responsible for the transformation of a metastable aragonite-dominated carbonate mud into an low-Mg calcite-dominated limestone.

(5) Thin-walled organic microfossils are reliable indicators for mechanical compaction. Their deformation provides a new, complementary tool for compaction assessment in fine-grained carbonates. The main advantage of this method is its applicability to limestones where other compaction indicators are absent.

(6) The dissolution of aragonite and the reprecipitation as microspar are probably located close to each other. Possible source layers of the carbonate cement are the compacted layers that are intercalated in the cemented, uncompacted layers. Textural similarities to limestone-marl alternations suggest that the formation of the examined alternations of cemented and uncemented layers may represent a "clay-free" equivalent of limestone-marl alternations. This is supported by the passive enrichment of palynomorphs and organic carbon in the uncemented layers. Sedimentary differences as the sole cause for the diagenetic differentiation were not supported by the data of the present study.

6 Outlook

The present study on Pliocene periplatform carbonates from the leeward slope of Great Bahama Bank (core CLINO) once more verifies the notion of highstand shedding of flat-topped carbonate platforms. Highstand shedding, however, is clearly observed only in the Upper Pliocene sediments, that have been deposited when a steep-sided morphology prevailed. The Lower Pliocene slope sediments studied were deposited when Great Bahama Bank exhibited a distally-steepened ramp morphology. In these sediments, the sea-level record was found to be much more difficult to decipher. This difference partly is accounted to the gently dipping morphology that defines a ramp. This morphology allows for a lateral shift of the facies belts instead of the abrupt cessation of carbonate production observed on flat-topped platforms during sea-level falls. Nevertheless, the pattern found also partly originated from the particular situation of Great Bahama Bank in the Lower Pliocene, when the high-frequency sea-level fluctuations were superimposed by an overall sea-level rise. Clearly, the signatures found in the present study can not be taken as representative for carbonate ramps in general. Following the work of Burchette and Wright (1992), further comparisons between different ramps, of which several are well-studied (e.g. Elrick et al., 1991; Elrick and Read, 1991; Bachmann et al., 1996), might lead to an understanding of the general sedimentation patterns.

A spatially extended examination of Great Bahama Bank has been initiated during ODP Leg 166 (Eberli et al., 1996). Drill sites were located downslope of the Bahamas Drilling Project (CLINO and UNDA) on the continuation of the Western Line (Figs. 3 and 4). Numerous studies on the cores recovered during the Leg 166 campaign are presently being carried out. These studies will lead to a better understanding of the areally extensive carbonate platform and of the effects of the morphologic evolution that have been examined in the present study.

A perspective that turned out valuable in this study is the combined consideration of sedimentologic and palynologic findings. The palynomorph associations from the samples from CLINO have been an important tool not only in assessing the environmental conditions present on the platform top and in confirming the subaerial exposure of the platform top and its colonization by terrestrial plants. Also, palynomorphs proved to be valuable in assessing diagenetic alterations of the fine-grained carbonates. Dinoflagellate cysts can be employed as compaction indicators, and palynomorphs, acting as insolubles in these pure carbonates, can indicate compaction by their passive enrichment. More work in this direction could establish the routine usage of palynomorphs in sedimentology.

The ultrafacies concept, formulated by Keupp (1977), was found appropriate in describing the diagenetic alterations of the periplatform carbonates. Especially in the fine-grained carbonates, ultrafacies definitions led to a description of the process of diagenetic differentiation in cemented and compacted layers that presumably will continue to differentiate during further diagenesis. The assumption, that the alternations observed can be understood as pure-carbonate equivalents to many diagenetically induced or enforced limestone-marl alternations, offers a new point of view for the re-examination of similar calcareous successions.

Furthermore, the ultrafacies concept as applied here potentially could supplement the examination of petrophysical properties of carbonates. As, for example, Biddle et al. (1992), Anselmetti and Eberli (1993-a, 1993-b, 1997), Kenter and Ivanov (1995), and Kenter et al. (1997) have shown, limestones show an acoustic behavior that deviates from that of siliciclastic rocks. While in the latter, acoustic impedance is predominantly a function of mineralogy and porosity, it is the type of porosity (intra- or intercrystallin, moldic, etc.) that is decisive for the acoustic impedance in limestones. This has been proven by light microscopic investigations of Anselmetti and Eberli (in press). The SEM examinations undertaken within the scope of the present study imply that there might be also a clear correlation between sonic velocity and ultrafacies. Systematic studies are necessary to gain a thorough understanding of carbonate diagenesis on a small (ultrafacies) scale in connection with physical properties, and finally with seismic interpretation.

References

Abreu V S, Anderson J B. (in press) Glacial Eustacy During the Cenozoic: Sequence Stratigraphic Implications. AAPG Bulletin

Abreu V S, Haddad G A (in press) Glacioeustatic fluctuations: the mechanism linking isotope stratigraphy and sequence stratigraphy from the Oligocene to middle Miocene. In: De Graciansky P-C, Hardenbol J, Jacquin T, Vail P R, Farley M B (eds) Sequence Stratigraphy of European Basins. (SEPM Special Publication, vol 59)

Adams J E, Rhodes M L (1960) Dolomitization by seepage refluxion. AAPG Bulletin 44: 1912-1920

Adey W H, Macintyre I G (1973) Crustose coralline algae: a re-evaluation in the geological sciences. Geological Society of America Bulletin 84: 883-904

Agassiz A (1894) A reconnaissance of the Bahamas and of the elevated reefs of Cuba in the steam yacht Wild Duck, January - April, 1893. Bull. Mus. Comp. Zoology Harvard Coll. 26: 1-203

Ahr W M (1973) The carbonate ramp: an alternative to the shelf model. Gulf Coast Association of Geological Societies New Orleans Transactions. 23: 221-225

Allan J R, Matthews R K (1982) Isotope signatures associated with early meteoric diagenesis. Sedimentology 29: 797-817

Anati D A, Gat J R (1989) Restricted marine basins and marginal sea environments. In: Fritz P, Fontes J C (eds) Handbook of Environmental Isotope Geochemistry. (vol 1). Elsevier, Amsterdam, pp 29-73

Andrews J E, Shepard F P, Hurley R J (1970) Great Bahama Canyon. Geological Society of America Bulletin 81: 1061-1078

Anselmetti F S, Eberli G P (1993-a) Controls on sonic velocity in carbonates. Pure and Applied Geophysics 141/2-4: 287-323

Anselmetti F S, Eberli G P (1993-b) Sonic velocity in carbonates and its correlation with depositional lithology and diagenesis. Internal Publication of the University of Miami

Anselmetti F S, Eberli G P (1997) Sonic velocity in carbonate sediments and rocks. In: Palaz, I, Marfurt K J (eds) Carbonate Seismology (SEG Geophysical Developments Series, vol). pp 53-74

Anselmetti F S, Eberli G P, in press, Sonic velocity in carbonates - a combined product of depositional lithology and diagenetic alteration. In: Ginsburg R N (ed) Ground Truthing Seismic Stratigraphy of a Prograding Carbonate Platform Margin: Neogene, Great Bahama Bank. (SEPM Contributions in Sedimentology)

Atkinson M, Smith S V, Stroup E D (1981) Circulation in Enewetak lagoon. Limnology and Oceanography 26: 1074-1083

Austin J A, Schlager W, Palmer A A. et al. (eds) (1986) Proceedings of the Ocean Drilling Program, Leg 101 (Initial Reports, vol. 101), Texas A&M University, College Station

Austin J A, Schlager W, Comet P A et al. (eds) (1988) Proceedings of the Ocean Drilling Program, Leg 101. (Scientific Results, vol 101).

Bachmann M, Bandel K, Kuss J, Willems, H (1996) Sedimentary Processes and Intertethyal Comparisons of Two Early/Late Cretaceous Carbonate Ramp Systems (NE-Africa and Spain). In: Reitner J, Neuweiler F, Gunkel F (eds) Global and Regional Controls on Biogenic Sedimentation. II Cretaceous Sedimentation. (Göttinger Arb. Geol. Paläont., vol Sb2). pp 151-163

Bachmann M, Willems H (1996) High-frequency cycles in the upper Aptian carbonates of the Organyà basin, NE Spain. Geologische Rundschau 85: 586-605

Baker P A, Bloomer S H (1988) The origin of celestite in deep-sea sediments. Geochimica et Cosmochimica Acta 52: 335-340

Ball M M (1967) Carbonate sand bodies of Florida and the Bahamas. Journal of Sedimentary Petrology 37: 556-591

Ball M M (1972) Exploration methods for stratigraphic traps in carbonate rocks. In: King R E (ed) Stratigraphic Oil and Gas Fields - Classification, Exploration Methods and Case Histories. (AAPG Memoir, vol 16). pp 64-81

Ball M M, Harrison C G A, Hurley R J, Leist C E (1969) Bathymetry in the vicinity of the northeastern scarp of the Great Bahama Bank and Exuma Sound. Bulletin of Marin Science, Gulf and Caribbean 19: 243-252

Bandel K, Hemleben C (1975) Anorganisches Kristallwachstum bei Mollusken. Paläontologische Zeitschrift 49: 293-320

Bathurst R G C (1966) Boring algae, micrite envelopes and lithification of molluscan biosparites. Journal of Geology 5: 15-32

Bathurst R G C (1970) Problems of lithification in carbonate muds. Proceedings of the Geologist's Association 81/3: 429-440

Bathurst R G C (1971) Carbonate Sediments and Their Diagenesis. (Developments in Sedimentology, vol 2). Elsevier, Amsterdam

Bathurst R C G (1975) Carbonate Sediments and Their Diagenesis. (Developments in Sedimentology, vol 12, 2nd edn). Elsevier, Amsterdam

Bathurst R G C (1986) Carbonate diagenesis and reservoir development: Conservation, destruction and creation of pores. In: Warme J E, Shanley K W (eds) Carbonate depositional environments, modern and ancient; part 5, Diagenesis I. (Colorado Scool of Mines Quaterly, vol 81/4) pp 1-25

Bathurst R G C (1993) Microfabrics in Carbonate Diagenesis: A Critical Look at Forty Years in Research. In: Rezak R, Lavoie D L (eds) Carbonate Microfabrics. Springer, Berlin Heidelberg New York, pp 3-14

Bausch W M (1997) Die Flexibilität der Kalk/"Mergel"-Grenze und ihre Berechenbarkeit. Zeitschrift der Deutschen Geologischen Gesellschaft 148: 247-258

Beach D K (1982) Depositional and Diagenetic History of Pliocene-Pleistocene Carbonates of Northwestern Great Bahama Bank; Evolution of a Carbonate Platform. Dissertation, University of Miami

Beach D K, Ginsburg R N (1980) Facies succession of Pliocene-Pleistocene Carbonates, Northwestern Great Bahama Bank. AAPG Bulletin 64: 1634-1642

Bhattacharyya A, Friedman G M (1979) Experimental compaction of ooids and lime mud and its implication for lithification during burial. Journal of Sedimentary Petrology 49: 1279-1286

Bibliographisches Institut Mannheim (1974) Meyers Großes Handlexikon. 11. edn. Meyers Lexikon Verlag, Mannheim

Biddle K T, Schlager W, Rudolph K W, Bush T L (1992) Seismic model of a progradational carbonate platform, Picco di Vallandro, the Dolomites, northern Italy. AAPG Bulletin 76: 14-30

Black M (1933) The algal sediments of Andros Island, Bahamas. Philosophical Transactions of the Royal Society London, Series B 222: 165-192

Blomeier D (1997) Evolution einer unterjurassischen Karbonatplattform: Sequenzstratigraphie und Ablagerungsbedingungen am Hochplateau des Jbel Bou Dahar (Hoher Atlas, Marokko). Dissertation, Christian-Albrechts-Universität Kiel

Boardman M R, Neumann A C (1984) Sources of periplatform carbonates: Northwest Providence Channel, Bahamas. Journal of Sedimentary Petrology 54: 1110-1123

Boardman M R, Neumann C A, Baker P A, Dulin L A, Kenter R J, Hunter G E, Kiefer K B (1986) Banktop responses to Quaternary fluctuations in sea level recorded in periplatform sediments. Geology 14: 28-31

Bond G, Lotti R (1995) Iceberg Discharges into the North Atlantic on Millenial Time Scales During the Last Glaciation. Science 267: 1005-1010

Bosellini A (1984) Progradation geometries of carbonate platforms: examples from the Triassic of the Dolomites, northern Italy. Journal of Sedimentology 31: 1-24

Bosellini A, Ginsburg R N (1971) Form and internal structure of recent algal nodules (rhodolites) from Bermuda. Journal of Geology 7: 669-682

Bosence D W J (1977) Ecological studies on two carbonate sediment-producing algae. In: Flügel E (ed) Fossil Algae. Springer, Berlin Heidelberg New York, pp 270-278

Bourrouilh-le-Jan F G (1980) Hydrologie des nappes d'eau superficielles d'Ile Andros, Bahama: Dolomitization et diagenese de plaine d'estran en pays tropical humide. Bulletin de Centre du Recherche, Exploration et Production, Elf-Aquitaine 4: 661-707

Brachert T C, Dullo W-C (1991) Laminar micrite crusts and associated foreslope processes, Red Sea. Journal of Sedimentary Petrology 61: 354-363

Brachert T C, Dullo W-C (1994) Micrite Crusts on Ladinian Foreslopes of the Dolomites Seen in the Light of a Modern Scenario from the Red Sea. Abhandlungen der Geologischen Bundesanstalt, Wien 50: 57-68

Bradford M R, Wall D A (1984) The distribution of Recent organic-walled dinoflagellate cysts in the Persian Gulf, Gulf of Oman and Northwestern Arabian Sea. Paleontographica B 192: 16-84

Broecker W S, Takahashi T (1966) Calcium carbonate precipitation on the Bahama Bank. Journal of Geophysical Research 71: 1575-1602

Bullard S C, Everett J, Smith A (1965) The fit of the continents around the Atlantic. Symposium on Continental Drift. London Philosphical Transactions of the Royal Society London Ser. A 258: 41-51

Burchette T P, Britton S R (1985) Carbonate facies analysis in the exploration for hydrocarbons: a case-study from the Cretaceous of the Middle East. In: Brenchley P J, Williams B P J (eds) Sedimentology - recent developments and applied aspects. Blackwell Scientific Publications, Oxford, pp 311-338

Burchette T P, Wright V P, 1992) Carbonate ramp depositional systems. Sedimentary Geology 79: 3-57

Canfield D E, Raiswell R (1991) Carbonate Precipitation and Dissolution. In: Allison P A, Briggs D E G (eds) Taphonomy: Releasing the data locked in the fossil record. Plenum Press, New York, pp 411-453

Carballo J D, Land L S, Miser D E (1987) Holocene dolomitization of supratidal sediments by active tidal pumping, Sugarloaf Key, Florida. Journal of Sedimentary Petrology 57: 153-156

Carlson W.D (1983) The polymorphs of $CaCO_3$ and the aragonite-calcite transformation. In: Reeder R J (ed) Carbonates: Mineralogy and Geochemistry. (Reviews in Mineralogy, Mineralogical Society of America, vol 11). pp 190-226

Cartwright R A (1985) Provenance and sedimentology of carbonate turbidites from two deep-sea fans, Bahamas. M.S. Thesis, University of Miami

Chatellier J.-Y (1988) Carboniferous carbonate ramp, the Banff Formation, Alberta, Canada. Bulletin des Centres de Recherches Exploration-Production Elf-Aquitaine 12: 569-599

Chayes F (1956) Petrographic model analysis. John Wiley & Sons, New York

Chilingar G V, Bissell H J, Wolf K H (1967) Diagenesis of carbonate rocks. In: Larsen G, Chilingar G V (eds) Diagenesis in sediments. (Developments in Sedimentology 8). Elsevier, Amsterdam. pp 179-322

Cloud P E (1962) Environment of Calcium Carbonate Deposition west of Andros Island, Bahamas. (U.S. Geological Survey Professional Papers, vol 350)

Constantz B R (1986) The primary surface of corals and variations in their susceptibility to diagenesis. In: Schroeder J H, Purser B H (eds) Reef diagenesis. Springer, Berlin Heidelberg New York, 53-76 pp

Correns C W (1949) Einführung in die Mineralogie (Kristallographie und Petrologie). Springer, Berlin Heidelberg New York

Craig H (1957) Isotopic standards for carbon and oxygen and correction factors for mass spectrometric analysis of carbon dioxide. Geochimica et Cosmochimica Acta 12: 133-149

Craton M (1992) A History of the Bahamas. (4. edn). San Salvador Press, Waterloo/Ontario

Crevello P D (1991) High-frequency carbonate cycles and stacking patterns: Interplay of orbital forcing and subsidence on Lower Jurassic rift platforms, High Atlas, Morocco. Kansas Geological Survey Bulletin 233: 207-230

Cutler W G (1983) Stratigraphy and sedimentology of the Upper Devonian Grosmont Formation, northern Alberta. Bulletin of Canadian Petroleum Geology 31: 282-235

Dansgaard W, Clausen H B, Gundestrup N S, Hammer C U, Johnsen S F, Kristinsdottir P M, Reeh N (1982) A new Greenland deep ice core. Science 218, p 1273-1277

Dansgaard W, Johnsen S J, Clausen H B, Dahl J D, Gundestrup N S, Hammer C U, Hvidberg C S, Steffensen J P, Sveinbjornsdottir A E, Jouzel J, Bond G (1993) Evidence for general instability of past climate from a 250-kyr ice-core record, Nature 364: 218-220

Davis J C (1973) Statistics and Data Analysis in Geology. John Wiley & Sons, New York

Dietrich G, Kalle K, Krauss W, Siedler G (1975) Allgemeine Meereskunde. (3 edn). Gebrüder Bornträger, Berlin Stuttgart

Dietz R S, Holden J C (1973) Geotectonic evolution and subsidence of Bahama Platform, Reply. Geological Society of America Bulletin 84: 3477-3482

Dietz R S, Holden J C, Sproll W P, 1970) Geotectonic evolution and subsidence of Bahama platform. Geological Society of America Bulletin 81: 1915-1927

Dix G R (1989) High-energy, inner shelf carbonate facies along a tide-dominated non-rimmed margin, northwestern Australia. Marine Geology 89: 347-362

Dix G R, Mullins H T (1988-a) Rapid burial diagenesis of deep-water carbonates: Exuma Sound, Bahamas. Geology 16: 680-683

Dix G R, Mullins H T (1988-b) A regional perspective of shallow burial diagenesis of deep-water periplatform carbonates from the northern Bahamas. In: Austin J A, Schlager W, Comet P A et al. (eds) Proceedings of the Ocean Drilling Program, Leg 101 (Scientific Results, vol 101) Texas A&M University, College Station, pp 279-304

Dix G R, Mullins H T (1992) Shallow burial diagenesis of deep-water carbonates, Northern Bahamas. Geological Society of America Bulletin 104: 305-315

Doran E (1955) Land forms of the southeastern Bahamas. (University of Texas Publications, vol 5509)

Dravis J.J (1977) Holocene sedimentary depositional environments on Eleuthera Bank, Bahamas. M.S. Thesis, University of Miami

Dravis J J (1996) Rapidity of freshwater calcite cementation - implications for carbonate diagenesis and sequence stratigraphy. Sedimentary Geology 107: 1-10

Droxler A W (1985) Last deglaciation in the Bahamas: a dissolution record from variations of aragonite content? In: Sundquist E T, Broecker W S (eds.): The carbon cycle and atmospheric CO_2: natural variations, Archean to Present. (Geophysical Monograph., vol 32). American Geophysical Union, Washington D.C., pp 195-207

Droxler A W, Bruce D, Sager W W, Watkins D K (1988) Pliocene-Pleistocene variations in aragonite content and planktonic oxygen isotope record in Bahamian periplatform ooze, Hole 633A. In: Austin J A, Schlager W, Comet P A et al. (eds) Proceedings of the Ocean Drilling Program, Leg 101. (Scientific Report, vol 101). Texas A&M University, College Station, pp 221-244

Droxler A W, Haddad G A, Mucciarone D A, Cullen J L (1990) Pliocene-Pleistocene aragonite cyclic variations in holes 714A and 716B (the Maldives) compared with hole 633A (the Bahamas): records of climate-induced $CaCO_3$ preservation at intermediate water depths. In: Duncan R A, Backman J, Peterson L C et al. (eds) Proceedings of the Ocean Drilling Program, Leg 115. (Scientific Results, vol 115). Texas A&M University, College Station, pp 539-577

Droxler A W, Morse J W, Glaser K S, Haddad G A, Baker P A (1991) Surface sediment carbonate mineralogy and water column chemistry: Nicaragua Rise versus the Bahamas. Marine Geology 100: 277-289

Droxler A W, Schlager W (1985) Glacial versus interglacial sedimentation rates and turbidite frequency in the Bahamas. Geology 13: 799-802

Droxler A W, Schlager W, Whallon C C (1983) Quaternary aragonite cycles and oxygen-isotope record in Bahamian Carbonate ooze. Geology 11: 235-239

Dullo W.-C (1983) Fossildiagenese im miozänen Leitha-Kalk der Parathetys von Österreich: Ein Beispiel der Faunenverschiebungen durch Diageneseunterschiede. Facies 8: 1-112

Dullo W.-C (1984) Progressive diagenetic sequence of aragonite structures: Pleistocene coral reefs and their modern counterparts on the eastern Red Sea coast, Saudi Arabia. Palaeontographica Americana 54: 254-260

Dullo W.-C (1986) Variations in diagenetic sequences: An example from Pleistocene coral reefs, Red Sea, Saudi Arabia. In: Schroeder J H, Purser B H (eds) Reef diagenesis. Springer, Berlin Heidelberg New York, pp 77-90

Dullo W.-C (1990) Facies, Fossil Record and Age of Pleistocene Reefs from Red Sea (Saudi Arabia). Facies 22: 1-47

Dullo W.-C, Camoin G F, Blomeier D, Colonna M, Eisenhauer A, Faure G, Casanova J, Thomassin B A (in press), Morphology and Sediments of the Foreslopes of Mayotte, Comoro Islands: Direct Obseravations from Submercible. In: Camoin G F, Bergerson D (eds) Reefs and Carbonate Platforms of the Indian Ocean and Pacific. (International Association of Sedimentologists Special Publication)

Dullo W.-C, Jado A.R (1984) Facies, zonation patterns and diagenesis of pleistocene reefs on the eastern Red Sea coast. Symposium on Coral Reef Environment of the Red Sea, Jeddah, pp 254-275

Dullo W-C, Moussavian E, Brachert T C (1990) The foralgal crust facies of the deeper forereefs in the Red Sea. Géobios 23: 261-281

Eberli G P, Ginsburg R N (1987) Segmentation and coalescene of Cenozoic carbonate platforms, northwestern Great Bahama Bank. Geology 15: 75-79

Eberli G P, Ginsburg R N (1989) Cenozoic progradation of northwestern Great Bahama Bank, a record of lateral platform growth and sea-level fluctuations. In: Crevello P D, Wilson J L, Sarg J F, Read J F (eds) Controls on carbonate platform and basin development. (SEPM Special Publication, vol 44). pp 339-351

Eberli G P, Kendall C G S C, Moore P, Whittle G L, Cannon R (1994) Testing a Seismic Interpretation of Great Bahama Bank with a Computer Simulation. AAPG Bulletin 78: 981-1004

Eberli G P, Kenter J A M, McNeill D F, Ginsburg R N, Swart P K, Melim L A (in press) Facies, Diagenesis and Timing of Prograding Seismic Sequences on Western Great Bahama Bank. In: Ginsburg R N (ed) Ground Truthing Seismic Stratigraphy of a Prograding Carbonate Platform Margin: Neogene, Great Bahama Bank. (SEPM Contributions in Sedimentology)

Eberli G P, Swart P K, Malone M (1996) Leg 166 Preliminary Report. (Ocean Drilling Program Preliminary Report, vol. 66). Texas A&M University, College Station

Eberli G P, Swart P K, McNeill D F, Kenter J A M, Anselmetti F S, Melim L A, Ginsburg R N (1997) A Synopsis of the Bahama Drilling Project: Results from two Deep Core Borings drilled on the Great Bahama Bank. In: Eberli G P, Swart P K, Malone, M. J. et al. (eds) Ocean Drilling Program, Leg 166 (Scientific Results, vol 166). pp 23-41

Eder W (1982) Diagenetic redistribution of carbonate, a process in forming limestone-marl alternations (Devonian and Carboniferous, Rheinisches Schiefergebirge, W. Germany). In: Einsele G, Seilacher A (eds.): Cyclic and Event Stratification. Springer, Berlin Heidelberg New York, pp 98-112

Edwards L E, Andrle V A S (1992) Distribution of selected dinoflagellate cysts in modern marine sediments. In: Head M J, Wrenn J H (eds) Neogene and Quaternary Dinoflagellate cysts and Acritarchs. American Asociation of Stratigraphic Palynologists Foundation, College Station, Texas, pp 259-288

Einsele G (1982) Limestone-marl cycles (periodites): diagnosis, significance, causes - a review. In: Einsele G, Seilacher A (eds) Cyclic and event stratification. Springer, Berlin Heidelberg New York, pp 8-53

Einsele G, Ricken W (1991) Limestone-Marl Alternations - an Overview. In: Einsele G, Ricken W, Seilacher A (eds) Cycles and events in stratigraphy. Springer, Berlin Heidelberg New York, pp 23-47

Elrick M, Read J F (1991) Cyclic ramp-to-basin carbonate deposits, Lower Mississippian, Wyoming and Montana: a combined field and computer modeling study. Journal of Sedimentary Petrology 61: 1194-1224

Elrick M, Read J F, Coruh C (1991) Short-term paleoclimatic fluctuations expressed in lower Mississippian ramp-slope deposits, southwestern Montana. Geology 19: 799-802

Emiliani C (1965) Precipitous continental slopes and considerations on the transitional crust. Science 147: 145-148

Enos P,1974) Map of surface sediment facies of the Florida-Bahama Plateau. (Geological Society of America Map Series, MC-5 no 4)

Enos P, 1983) Shelf Environment. In: Scholle, P A., Bebout, D G, Moore, C H (eds) Carbonate depositional environments. (AAPG Memoir, vol 33). pp 267-295

Enos P, Sawatsky L H (1981) Pore networks in Holocene carbonate sediments. Journal of Sedimentary Petrology 51: 961-985

Ericsson D B, Ewing M, Heezen B (1952) Turbidity currents and sediments in the North Atlantic. AAPG Bulletin 36: 489-511

Everts A J W (1991) Interpreting compositional variations of calciturbidites in relation to platform-stratigrapy: an example from the Paleogene of SE Spain. Sedimentary Geology 71: 231-242

Everts A J W (1994) Carbonate Sequence Stratigraphy of the Vercors (French Alps) and its Bearing on the Cretaceous. Dissertation, Vrije Universiteit Amsterdam

Everts A J W, Reijmer J J G (1995) Clinoform composition and margin geometries of a Lower Cretaceous carbonate platform (Vercors, SE France). Palaeogeography-Palaeoclimatology-Palaeoecology 119: 19-33

Ferronsky V I, Brezgunov V S (1989) Stable isotopes and ocean mixing. In: Fritz P, Fontes J C (eds) Handbook of Environmental Isotope Geochemistry. (vol 1). Elsevier, Amsterdam, pp 21-47

Fischer A G, Honjo S, Garrison R E (1967) Electron Micrographs of Limestones and their Nannofossils. Princeton University Press, Princeton

Flajs G (1977) Die Ultrastrukturen des Kalkalgenskeletts. Palaeontographica 160 Abt. B: 69-128

Flügel E (1967) Elektronenmikroskopische Untersuchungen an mikritischen Kalken. Geologische Rundschau 56: 341-358

Flügel E (1978) Mikrofazielle Untersuchungsmethoden von Kalken. Springer, Berlin Heidelberg New York

Flügel E (1982) Microfacies Analysis of Limestones. Springer, Berlin Heidelberg New York

Folk R L (1959) Practical petrographic classification of limestones. AAPG Bulletin 43: 1-38

Folk R L (1965) Some aspects of recrystallization in ancient limestones. In: Pray L C, Murray R C (eds) Dolomitization and Limestone Diagenesis. (SEPM Special Publication, vol 13). pp 14-48

Folk R L (1974) The natural history of crystalline calcium carbonate: effect of magnesium content and salinity. Journal of Sedimentary Petrology 44: 141-153

Folk R L, Land L S (1975) Mg/Ca Ratio and Salinity: Two controls over Crystallization of Dolomite. AAPG Bulletin 59: 60-68

Freile D, Milliman J D, Hillis L (1995) Leeward bank margin Halimeda meadows and draperies and their sedimentary importance on the western Great Bahama Bank slope. Coral Reefs 14: 27-33

Friedman G M (1964) Early diagenesis and lithification in carbonate sediments. Journal of Sedimentary Petrology 34: 777-813

Friedman G M (1965) Occurrence and stability relationships of aragonite, high-magnesian calcite and low-magnesium calcite under deep-sea conditions. Geological Society of America Bulletin 76: 1191-1196

Fruth L S J, Orme G R, Donath F A (1966) Experimental compaction effects in carbonate sediments. Journal of Sedimentary Petrology 36: 747-754

Gabb W M (1873) Topography and geology of Santo Domingo. Am. Philos. Soc. Trans., n. ser. 15: 49-259

Gibson T B, Schlee J (1967) Sediments and fossiliferous rocks from the eastern side of the Tongue of the Ocean, Bahamas. Deep-Sea Research 14: 691-702

Ginsburg R N (1956) Environmental relationships of grain size and constituent particles in some South Florida carbonate sediments. AAPG Bulletin 40: 2384-2427

Ginsburg R N (in press-a) Ground Truthing Seismic Stratigraphy of a Prograding Carbonate Platform Margin: Neogene, Great Bahama Bank. (SEPM Contributions in Sedimentology)

Ginsburg R N (in press-b) The Bahamas Drilling Project: Summaries of Background and Acquisition of Cores and Logs. In: Ginsburg R N (ed) Ground Truthing Seismic Stratigraphy of a Prograding Carbonate Platform Margin: Neogene, Great Bahama Bank. (SEPM Contributions in Sedimentology)

Ginsburg R N, Eberli G P, McNeill D F, Swart P K, Kenter J A M, Anselmetti F, Melim L, Kievman C, Warzeski R (in press) Calibration of seismic stratigraphy, burial diagenesis and platform anatomy of Great Bahama Bank: results of drilling from an alternate platform. In: Ginsburg R N (ed) Ground Truthing Seismic Stratigraphy of a Prograding Carbonate Platform Margin: Neogene, Great Bahama Bank. (SEPM Contributions in Sedimentology)

Ginsburg R N, Harris P M, Eberli G P, Swart P K (1991) The growth potential of a bypass margin, Great Bahama Bank. Journal of Sedimentary Petrology 61: 976-987

Glaser K S, Droxler A W 1993) Controls and development of late Quaternary periplatform carbonate stratigraphy in Walton Basin (northeastern Nicaragua Rise, Caribbean Sea). Paleoceanography 8: 243-274

Glenn C R, Kronen J D Jr (1993) Origin and Significance of Late Pliocene Phosphatic Hardgrounds on the Queensland Plateau, Northeastern Australian Margin. In: McKenzie J A, Davies P J, Palmer-Julson A et al. (eds) Proceedings of the Ocean Drilling Program, Leg 133. (Scientific Results, vol 133). pp 525-534

Glockhoff C (1973) Geotectonic evolution and subsidence of Bahama platform: Discussion. Geological Society of America Bulletin 84: 3473-3476

Goldhammer R K, Dunn P A, Hardie L A (1990) Depositional cycles, composite sea-level changes, cycle stacking patterns and the hierarchy of stratigraphic forcing: examples from Alpine Triassic platform carbonates. Geological Society of America Bulletin 102: 535-562

Goldhammer R K, Harris M T (1989) Eustatic controls on the stratigraphy and geometry of the Latemar Buildup (Middle Triassic), the Dolomite of northern Italy. In: Crevello P D, Wilson J L, Sarg J F, Read J F (eds) Controls on carbonate platform and basin development. (SEPM Special Publication, vol 44). pp 323-338

Goldsmith J R, Graf D L (1958) Relation between lattice constrains and composition of the Ca-Mg carbonates. American Mineralogist 43: 84-101

Goldstein R H (1988) Cement Stratigraphy of Pennsylvanian Holder Formation, Sacramento Mountains, New Mexico. AAPG Bulletin 72: 425-438

Goldstein R H (1990) Petrographic and geochemical evidence for origin of paleospeleothems, New Mexico: Implications for the application of fluid inclusions to studies of diagenesis. Journal of Sedimentary Petrology 60: 282-292

Goldstein R H, Reynolds T J (1994) Systematics of Fluid Inclusions in Diagenetic Minerals. (SEPM Short Course Notes, vol 31)

Goren M (1979) Succession of benthic community on artificial substratum at Elat (Red Sea). Journal of Experimental Marine Biology and Ecology 38: 19-40

Graf D L, Goldsmith J R (1956) Some hydrothermal syntheses of dolomite and protodolomite. Geology 64: 173-186

Grammer G M, Ginsburg R N (1992) Highstand versus lowstand deposition on carbonate platform margins: insight from Quaternary foreslopes in the Bahamas. Marine Geology 103: 125-136

Grammer G M, Ginsburg R N, Harris P M (1993-a) Timing of deposition, diagenesis and failure of steep carbonate slopes in response to a high-amplitude/ high-resolution fluctuation in sea level,

Tongue of the Ocean, Bahamas. In: Loucks R G, Sarg J F (ed) Carbonate Sequence Stratigraphy. (AAPG Memoir, vol 57). pp 107-131

Grammer G M, Ginsburg R N, McNeill D F (1991) Morphology and development of modern carbonate foreslopes, Tongue of the Ocean, Bahamas. 12th Carribean Geological Conference, St. Croix U.S.V.I.. Miami Geological Society: 27-32

Grammer G M, Ginsburg R N, Swart P K, McNeill D F, Jull A J, Prezbindowski D R (1993-b) Rapid growth rates of syndepositional marine aragonite cements in steep marginal slope deposits, Bahamas and Belize. Journal of Sedimentary Petrology 63: 983-989

Gross M G (1964) Variations in the O^{18}/O^{16} and C^{13}/C^{12} ratios of diagenetically altered limestones in the Bermuda Islands. Journal of Geology 72: 170-194

Grossman E L, Ku T.-L (1986) Oxygen and carbon isotope fractionation in biogenic aragonite: temperature effects. Chemical Geology 59: 59-74

Gulbrandsen R A (1960) Method of X-Ray analysis for determining the ration of calcite to dolomite in mineral mixtures. U. S. Geological Survey Bulletin 111: 147-152

Gunatilaka A (1976) Thallophyte boring and micritization within skeletal sands from Connemara, Western Ireland. Journal of Sedimentary Petrology 46: 548-554

Haak A B, Schlager W (1989) Compositional variations in calciturbidites due to sea-level fluctuations, late Qaternary, Bahamas. Geologische Rundschau 78: 477-486

Halley R B (1987) Diagenesis 2. Burial diagenesis of carbonate rocks. Colorado School of Mines Quaterly 82: 1-15

Haq B U, Hardenbol J, Vail P R (1988) Mesozoic and Cenozoic chronostratigraphy and cycles of sea-level changes. In: Wilgus C K, Hastings B S, Kendall C G S C, Posamentier H W, Ross C A et al. (eds.): Sea-level changes: an integrated approach. (SEPM Special Publication, vol 42). pp 71-108

Harris M T (1994) The foreslope and toe-of-slope facies of the Middle Triassic Latemar Buildup (Dolomites, Northern Italy). Journal of Sedimentary Research 64: 132-145

Head M J, Westphal H (submitted) Subsurface marinepalynology and paleoenviroments of a Pliocene carbonate platform: the Clino Core, Bahamas. Micropaleontology

Head M J (1997) Thermophilic dinoflagellate assemblages from the mid Pliocene of eastern England. Journal of Paleontology 71: 165-193

Heinrich H (1988) Origin, consequence of cyclic ice rafting in the Northeast Atlantic Ocean during the past 130,000 years. Quaternary Research 29: 142-152

Hendry J P, Trewin N H, Fallick A E (1996) Low-Mg calcite marine cement in Cretaceous turbidites: origin, spatial distribution and relationship to sea-water chemistry. Sedimentology 43: 877-900

Herrera F de (1601) Historia general de las Indias occidentales

Hess H H (1933) Submerged river valleys of the Bahamas. American Geophysical Union Transactions (14th Anniver. Meeting): 168-170

Hess H H (1960) The origin of the Tongue of the Ocean and other great valleys of the Bahama Banks. 2nd Caribbean Geol. Conf., Mayaguez, Puerto Rico, pp 160-161

Hilgard E W (1871) On the geological history of the Gulf of Mexico. American Journal of Science, 3rd ser. 2: 391-404

Hilgard E W (1881) The later Tertiary of the Gulf of Mexico. American Journal of Science, 3rd ser. 22: 58-65

Hölemann J A (1994) Akkumulation von autochthonem und allochthonem organischen Material in den känozoischen Sedimenten der norwegischen See (ODP Leg 104). (Geomar Report, vol 33). Geomar, Kiel

Honjo S (1969) Study of fine grained carbonate matrix: sedimentation and diagenesis of "micrite". Paleontological Society of Japan Special Paper 14: 67-82

Hook J E, Golubic S, Milliman J D (1984) Micritic cement in microborings is not necessarily a shallow-water indicator. Journal of Sedimentary Petrology 54: 425-431

Hudson J D (1962) Pseudo-pleochroic Calcite in Recrystallized Shell-limestones. Geological Magazine 99: 492-500

Husseini S I, Matthews R K (1972) Distribution of high-magnesium calcite in lime muds of the Great Bahama Bank: Diagenetic implication. Journal of Sedimentary Petrology 42: 179-182

Illing L V (1954) Bahamian calcareous sands. AAPG Bulletin 38: 1-95

Illing M A (1950) The Mechnical Distribution of Recent Foraminifera in Bahama Bank Sediments. Ann. Mag. Nat. Hist. 12, ser. iii: 757-761

Illing M A (1952) Distribution of Certain Foraminifera within the Littoral Zone on the Bahama Banks. Ann. Mag. Nat. Hist. 12, ser. v: 275-285

James N P, Choquette P W (1983-a) Diagenesis-5. Limestones. Geoscience Canada 10: 159-161

James N P, Choquette P W (1983-b) Diagenesis-6. Limestones - The sea-floor diagenetic environment. Geoscience Canada 10: 162-180

James N P, Choquette P W (1984) Diagenesis-9: Limestones - the meteoric diagenetic environment. Geoscience Canada 11: 161-194

James N P, Ginsburg R N (1979) The Seaward Margin of Belize Barrier and Atoll Reefs. (International Association of Sedimentologists Special Publication, vol 3)

Joachimski M M (1994) Subaerial exposure and deposition of shallowing upward sequences: evidence from stable isotopes of Purbeckian peritidal carbonates (basal Cretaceous), Swiss and French Jura Mountains. Sedimentology 41: 805-824

Johnsen S J, Clausen H B, Dansgaard W, Fuhrer K, Gundestrup N, Hammer C U, Iversen P, Jouzel J, Stauffer B, Steffensen J P, 1992) Irregular glacial interstadials recorded in a new Greenland ice core. Nature 359: 311-313

Joint Committee on Powder Diffraction Standards (1970) Index of the Powder Diffraction File 1970. Philadelphia, PA

Kaufman J (1994) Numerical models of fluid flow in carbonate platforms: implications for dolomitization. Journal of Geophysical Research 64: 128-139

Kendall C G S C, Schlager W (1981) Carbonates and relative changes in sea level. Marine Geology 44: 181-212

Kennard J M, Southgate P N, Jackson M J, O'Brian ppE (1992) New sequence perspective on the Devonian reef complex and Frasnian-Famennian boundary, Canning Basin, Australia. Geology 20: 1135-1138

Kenter J A M, Ivanov M (1995) Parameters controlling acoustic properties of carbonate and siliciclastic sediments at sites 866 and 869. In: Winterer E L, Sager W W, Firth J V et al. (eds) Proceedings of the Ocean Drilling Program, Leg 143 (Scientific Results. vol 143). pp 287-303

Kenter J A M, Podladchikov F F, Reinders M, van der Gaast S, Fouke B W, Sonnenfeld M D (1997) Parameters controlling sonic velocities in a mixed carbonate siliciclastics Permian shelf-margin (upper San Andres Formation, Last Chance Canyon, New Mexico). Geophysics 62: 505-520

Kenter J A M, Ginsburg R N, Troelstra S R, in press) The Western Great Bahama Bank: Sea-level-driven sedimentation patterns on the slope and margin. In: Ginsburg R N (ed) Ground Truthing Seismic Stratigraphy of a Prograding Carbonate Platform Margin: Neogene, Great Bahama Bank. (SEPM Contributions in Sedimentology)

Keupp H (1977) Ultrafazies und Genese der Solnhofer Plattenkalke (Oberer Malm, südliche Frankenalb). (Natur und Mensch, vol 37)

Kier J S, Pilkey O H (1971) The influence of sea level changes on sediment carbonate mineralogy, Tongue of the Ocean, Bahamas. Marine Geology 11: 189-200

Kievman C M, Ginsburg R.N, in press) Pliocene to Pleistocene depositional history of the upper platform margin, northwest Great Bahama Bank. In: Ginsburg R N (ed) Ground Truthing Seismic Stratigraphy of a Prograding Carbonate Platform Margin: Neogene, Great Bahama Bank. (SEPM Contributions in Sedimentology)

King R H (1947) Sedimentation in Permian Castile Sea. AAPG Bulletin 31: 470-477

Kohout F A, Henry H R, Banks J E (1977) Hydrogeology related to the geothermal conditions of the Floridan Plateau. (Florida Bureau of Geology Special Publication, vol 21) pp 1-39

Land L S, Mackenzie F T, Gould S J (1967) Pleistocene history of Bermudas. Bulletin of the Geological Society of America 78: 993-1006

Lasemi Z, Boardman M R, Sandberg P A (1988) Cement origin of supratidal dolomite, Andros Island, Bahamas. Journal of Sedimentary Petrology 59: 249-257

Lasemi Z, Boardman M R, Sandberg P A (1990) New microtextural criterion for differentiation of compaction and early cementation in fine grained limestones. Geology 18: 370-373

Lasemi Z, Sandberg P A (1984) Transformation of aragonite-dominated lime muds to microcristalline limestones. Geology 12: 420-423

Lasemi Z., Sandberg P A (1993) Microfabric and Compositional Clues to Dominant Mud Mineralogy of Micrite Precursors. In: Rezak R, Lavoie D.L (eds) Carbonate Microfabrics. Springer, Berlin Heidelberg New York, pp 173-185

Le Pichon X, Fox P J (1971) Marginal zones, fracture zones and the early opening of the North Atlantic. Journal of Geophysical Research 76: 6294-6308

Lidz B H, Bralower T J (1994) Microfossil biostratigraphy of prograding Neogene platform margin carbonates, Bahamas: Age constraints and alternatives. Marine Micropaleontology 23: 265-344

Lidz B H, McNeill D F (1995-a) Deep-sea biostratigraphy of prograding platform margins (Neogene, Bahamas): key evidence linked to depositional rhythm. Marine Micropaleontology 25: 87-125

Lidz B H, McNeill D F (1995-b) Reworked Paleogene to early Neogene planktic foramininfera: implications of an intriguing distribution at a late Neogene prograding margin, Bahamas. Marine Micropaleontology 25: 221-268

Loeblich A R, Tappan H (1964) Sarcodina chiefly "Thecameobians" and Foraminiferida. In: Moore A R (ed) Treatise of Invertebrate Paleontology, Part C, Protista 2. University of Kansas and Geological Society of America, Lawrence, Kansas

Lohmann K C, Meyers W J (1977) Microdolomite inclusions in cloudy prismatic calcites: a proposed criterion for former high-Magnesium calcites. Journal for Sedimentary Geology 47: 1078-1088

Longman M W (1981) Carbonate diagenesis as a control on stratigraphic traps. AAPG Education Note Series 21: 1-159

Loreau J P (1982) Sédiments aragonitiques et leur genèse. (Mémoires du Muséum National d'Histoire Naturelle, Série C, vol Tome XLVII)

Lowenstam H A, Epstein S (1957) On the origin of sedimentary aragonite needles of the Great Bahama Bank. Journal of Geology 65: 364-375

Lumsden D N (1979) Discrepancy between thin-section and X-ray estimates of dolomite in limestone. Journal of Sedimentary Petrology 49: 429-436

Lynts G W (1970) Conceptual model of the Bahamian Platform for the last 135 million years. Nature 225: 1226-1228

Lynts G W, Judd J B, Stehmann C F (1973) Late Pleistocene history of Tongue of the Ocean, Bahamas. Geological Society of America Bulletin 84: 2665-2684

Macintyre I G, Reid R P (1992) Comment on the origin of aragonite needle mud, a picture is worth a thousand words. Journal of Sedimentary Petrology 62: 1095-1097

Major R P (1991) Cathodoluminescene in post Miocene carbonates In: Barker C E, Kopp O C, Luminescence microscopy and spectroscopy: qualitative and quantitative applications. (SEPM Short Course, vol 25). pp 149-154

Maliva R G (1995) Recurrent neomorphic and cement microtextures from different diagenetic environments, Quaternary to Late Neogene carbonates, Great Bahama Bank. Sedimentary Geology 97: 1-7

Matthews R K (1966) Genesis of recent lime mud in southern British Honduras. Journal of Sedimentary Petrology 36: 428-454

McNeill D F (1989) Magnetostratigraphic Dating and Magnetization of Cenozoic Platform Carbonates from the Bahamas. Dissertation, University of Miami

McNeill D F, Eberli G P, Lidz B H, Swart P K, Kenter J A M (in press) Chronostratigraphy of prograding carbonate platform margins: A record of dynamic slope sedimentation, Western Great Bahama Bank. In: Ginsburg R N (ed) Ground Truthing Seismic Stratigraphy of a

Prograding Carbonate Platform Margin: Neogene, Great Bahama Bank. (SEPM Contributions in Sedimentology)

McNeill D, Ginsburg R N, Chang S.-B R, Kirschvink J L (1988) Magnetostratigraphic dating of shallow-water carbonates from San Salvador, Bahamas. Geology 16: 8-12

Melim L A (1996) Limitations on lowstand meteoric diagenesis in the Pliocene-Pleistocene of Florida and Great Bahama Bank: Implications for eustatic sea-level models. Geology 24: 893-896

Melim L A, Anselmetti F S, Eberli G P (in press-a) The importance of pore type on permeability of neogene carbonates, Great Bahama Bank. In: Ginsburg R N (ed) Ground Truthing Seismic Stratigraphy of a Prograding Carbonate Platform Margin: Neogene, Great Bahama Bank. (SEPM Contributions in Sedimentology)

Melim L A, Swart P K, Maliva R G (1995) Meteoric-like fabrics forming in marine waters: Implications for the use of petrography to identify diagenetic environments. Geology 23: 755-758

Melim L A, Swart P K, Maliva R G (in press-b) Meteoric and Marine Burial Diagenesis in the Subsurface of the Great Bahama Bank. In: Ginsburg R N (ed) Ground Truthing Seismic Stratigraphy of a Prograding Carbonate Platform Margin: Neogene, Great Bahama Bank. (SEPM Contributions in Sedimentology)

Meyerhoff A A, Hatten C W (1974) Bahamas salient of North America: Tectonic framework, stratigraphy and petroleum potential. AAPG Bulletin 58: 1201-1239

Meyers W J (1978) Carbonate cements: Their regional distribution and interpretation in Mississippian limestones of southwestern New Mexico. Sedimentology 25: 371-400

Meyers W J (1980) Compaction in Mississippian skeletal limestones, southwestern New Mexico. Journal of Sedimentary Petrology 50: 457-474

Milliman J D (1967) The geomorphology and history of Hogsty Reef, a Bahamian atoll. Bulletin of Marine Science 17: 519-543

Milliman J D (1974) Marine Carbonates. Springer, Berlin Heidelberg New York

Milliman J D (1977) Role of Calcareous Algae in Atlantic Continental Margin Sedimentation. In: Flügel E (ed). Fossil Algae, Recent Results and Developments. Springer, Berlin Heidelberg New York, pp 232-247

Mitchum R M, Sangree J B, Vail P R, Wornardt W W (1994) Recognizing Sequences and Systems Tracts from Well Logs, Seismic Data, and Biostratigraphy: Examples from the Late Cenozoic of the Gulf of Mexico. In: Weimer P, Posamentier H W (eds) Siliciclastic Sequence Stratigraphy (AAPG Memoir, vol 58). pp 163-197

Mix A C, Le J, Shackleton N J (1995-a) Benthic foraminiferal stable isotope stratigraphy of Site 846: 0-1.8 Ma. In: Pisias N G, Mayer L A, Janecek T R et al. (eds) Proceedings of the Ocean Drilling Program, Leg 138. (Scientific Results, vol 138). pp 337-357

Mix A C, Pisias N G, Rugh W, Wilson J, Morey A, Hagelberg T K (1995-b) Benthic foraminifer stable isotope record from Site 849 (0-5 Ma): local and global climate changes. In: Pisias N G, Mayer L A, Janecek T R et al. (eds) Proceedings of the Ocean Drilling Program, Leg 138. (Scientific Results, vol 138) pp 337-357

Montaggioni L F (1988) Holocene reef growth history in mid-plate high volcanic islands. In: Choat J H, Barnes D, Boronitzka M A et al. (eds) Proceedings of the 6th international Coral Reef Symposium, Townsville/Australia. (vol 3). pp 455-460

Morse J W, Mackenzie F T (1990) Geochemistry of Sedimentary Carbonates. (Developments in Sedimentology, vol 48). Elsevier, New York

Morse J W, Mackenzie F T (1993) Geochemical constraints on $CaCO_3$ transport in subsurface sedimentary environments. Chemical Geology 105: 181-196

Morzadec-Kerfourn M T (1983) Interèt des kystes de dinoflagellés pour l'établissement de reconstitution paleogéographique: exèmple du Golfe de Gabes (Tunisie). Cahiers de Micropaleontologie 1983/4: 15-22

Morzadec-Kerfourn M T (1992) Estuarine dinoflagellate cysts among oceanic assemblages of Pleistocene deep-sea sediments from the West African margin and their paleoenvironmental significance. In: Head M J, Wrenn J H (eds) Neogene and Quaternary Dinoflagellate cysts and Acritarchs. American Asociation of Stratigraphic Palynologists Foundation, College Station, Texas, pp 133-146

Moshier S O (1989) Microporosity in micritic limestones: a review. Sedimentary Geology 63: 191-213

Mullins H T (1975) Stratigraphy and structure of Northeast Providence Channel, Bahamas and origin of the northwestern Bahama Platform. M.S. Thesis, Duke University, Durham

Mullins H T (1983) Modern carbonate slopes and basins of the Bahamas. In: Cook H E, Hine A C, Mullins H T (eds) Platform margin and deep water carbonates. (SEPM Short Course, vol 12). pp 4.1-4.138

Mullins H T (1986) Periplatform carbonates. Colorado School of Mines Quarterly 81: 1-63

Mullins H T, Land L S, Wise W W J, Seigel D I, Masters P M, Hinchey E J, Price K R (1985-a) Authigenic dolomite in Bahamian periplatform slope sediment. Geology 13: 292-295

Mullins H T, Lynts G W (1977) Origin of the northwestern Bahama Platform: review and interpretation. AAPG Bulletin 88: 1447-1461

Mullins H T, Wise W W J, Gradulski A F, Hinchey E J, Masters P M, Siegel D I (1985-b) Shallow subsurface diagenesis of Pleistocene periplatform ooze: Northern Bahamas. Sedimentology 32: 473-494

Munnecke A (1997) Bildung mikritischer Kalke im Silur auf Gotland. Courier Forschungsinstitut Senckenberg 198: 1-71

Munnecke A, Samtleben C (1996) The formation of micritic limestones and the development of limestone-marl alternations in the Silurian of Gotland, Sweden. Facies 34: 159-176

Munnecke A, Servais T 1996) Scanning Electron Microscopy of Polished, Slightly Etched Surfaces of Silurian Limestones from Gotland: a Method to Observe Acritarchs *in Situ*. Palynology 40: 163-176

Munnecke A, Westphal H, Reijmer J J G, Samtleben C (1997) Microspar development during early marine burial diagenesis: a comparison of Pliocene carbonates from the Bahamas with Silurian limestones from Gotland (Sweden). Sedimentology 44 :977-990

Murray J W (1973) Distribution and Ecology of Living Benthic Foraminiferids. Heinemann Educational Books, London

Nelson R J (1853) On the geology of the Bahamas and on coral formation generally. Quarterly Journal of the Geological Society of London 9: 200-215

Neumann A C (1965) Processes of recent carbonate sedimentation in Harrington Sound, Bermudas. Bulletin of Marine Sciences 15: 987-1035

Neumann A C, Land L S (1975) Lime mud deposition and calcareous algae in the Bight of Abaco, Bahamas: a budget. Journal of Sedimentary Petrology 59: 147-161

Newell N D (1955) Bahamian platforms. In: Poldervaart, A (ed) Crust of the Earth. (Geological Society of America Special Paper, vol 62). pp 303-316

Newell N D, Imbrie J (1955) Biogeological reconnaissance in the Bimini area, Great Bahama Bank. Transactions of the New York Academy of Sciences 18: 3-14

Newell N D, Rigby J K (1957) Geological studies in the Great Bahama Bank. In: R J Le Blanc and J G Breeding (eds) Regional Aspects of Carbonate Sedimentation. SEPM Special Publication 5: 15-79

Newell N D, Rigby J K, Whiteman A J, Bradley J S (1951) Shoal Water Geology and Environments, Eastern Andros Island, Bahamas. Bull. Amer. Mus. Nat. Hist 97: Art. 1

Palmer M S (1979) Holocene Facies Geometry of the Leeward Bank Margin, Tongue of the Ocean, Bahamas. M.S. Thesis Thesis, University of Miami

Paulus F J (1972) The geology of site 98 and the Bahama Platform. In: Hollister C D, Ewing J I, Habib D, Hathaway J C, Lancelot Y, Luterbacher H, Paulus F J, Poag C W, Wilcoxon J A, Worstell P, Kaneps A G (eds), Initial Reports of the Deep Sea Drilling Project. (vol 11). U. S. Government Printing Office, Washington D. C, pp 877-897

Pilkey O H, Rucker J (1966) Mineralogy of Tongue of the Ocean sediments. Journal of Marine Research 24: 276-285

Pingitore N E (1970) Diagenesis and porosity modification in Acropora palmata, Pleistocene of Barbados, West Indies. Journal of Sedimentary Petrology 40: 712-721

Pomar L (1993) High-Resolution Sequence Stratigraphy in Prograding Miocene Carbonates: Application to Seismic Interpretation. In: Loucks R G, Sarg J F (eds) Carbonate Sequence Stratigraphy. (AAPG Memoir, vol 57). pp 389-407

Posamentier H W, Allen G P, James D P, Tesson M (1992) Forced regressions in a sequence stratigraphic framework: concepts, examples and exploration significance. AAPG Bulletin 76: 1687-1709

Pray L C (1960) Compaction of calcilutites (abstract). Bulletin of the Geological Society of America 71: 1946

Purdy E G (1963) Recent calcium carbonate facies of the Great Bahama-Bank. Journal of Geology 71: 334-355

Queen J M (1978) Hydrology, sedimentology and ecology of pelleted carbonate muds, west of Andros Island, Bahamas. Dissertation, State University of New York, Stony Brook

Raiswell R (1988) Chemical model for the origin of minor limestone-shale cycles by anaerobic methane oxidation. Geology 16: 641-644

Read J F (1982) Carbonate platforms of passive (extensional) continental margins: types, characteristics and evolution. Tectonophysics 81: 195-212

Read J F (1985) Carbonate Platform Facies Models. AAPG Bulletin 69: 1-21

Reid R P, Macintyre I G, James N P, 1990) Internal precipitation of microcrystalline carbonate: a fundamental problem for sedimentologists. Sedimentary Geology 68: 162-170

Reid R P, Carey S N, Ross D R (1996) Late Quaternary sedimentation in the Lesser Antilles island arc. Geological Society of America Bulletin 108: 78-100

Reijmer J J G, Schlager W, Droxler A W (1988) Site 632: Pliocene-Pleistocene Sedimentation Cycles in a Bahamian Basin. In: Austin J A, Schlager W, Comet P A et al. (eds) Proceedings of the Ocean Drilling Program, Leg 101. (Scientific Results, vol 101). pp 213-220

Reijmer J J G (1991) Sea level and sedimentation on the flanks of carbonate platforms. Dissertation, Vrije Universiteit, Amsterdam

Reijmer J J G, Everaars J S L (1991) Carbonate platform facies reflected in carbonate basin facies (Triassic, Northern Calcareous Alps, Austria). Facies 25: 253-278

Reijmer J J G, Schlager W, Bosscher H, Beets C J, McNeill D F (1992) Pliocene/Pleistocene platform facies transition recorded in calciturbidites (Exuma Sound, Bahamas). Sedimentary Geology 78: 171-179

Reijmer J J G, Ten Kate W G H, Sprenger A, Schlager W (1991) Calciturbidite composition related to exposure and flooding of a carbonate platform (Triassic, Eastern Alps). Sedimentology 38: 1059-1074

Rich J L (1951) Three critical environments of deposition and criteria for recognition of rocks deposited in each of them. Bulletin of the Geological Society of America 62:1-20

Ricken W (1986) Diagenetic Bedding: A Model for Limestone-Marl Alternations. (Lecture Notes in Earth Science, vol 6). Springer, Berlin Heidelberg New York

Ricken W (1987) The carbonate compaction law: a new tool. Sedimentology 34: 1-14

Ricken W (1993) Sedimentation as a Three-Component System. (Lecture Notes in Earth Science, vol 51). Springer, Berlin Heidelberg New York

Ricken W, Eder W (1991) Diagenetic modification of calcareous beds - an overview. In: Einsele G, Ricken W, Seilacher A (eds) Cycles and events in stratigraphy. Springer, Berlin Heidelberg New York, pp 430-449

Robbins L L, Blackwelder P L (1992) Biochemical and ultrastructural evidence for the origin of whitings: A biologically induced calcium carbonate precipitation mechanism. Geology 20: 464-468

Rock N M S (1988) Numerical Geology (Lecture Notes in Earth Science, vol 18) Springer, Berlin Heidelberg New York

Rose P R, Lidz B (1977) Diagnostic foraminiferal assemblages of shallow-water modern environments: south Florida and the Bahamas. (Sedimenta, vol VI). University of Miami

Royse C F J, Wadell J S, Petersen L E (1971) X-ray determination of calcite-dolomite: an evaluation. Journal of Sedimentary Petrology 41: 483-488

Rucker J B (1968) Carbonate mineralogy of sediments of Exuma Sound, Bahamas. Journal of Sedimentary Petrology 38: 68-72

Rush P F, Chafetz H S (1990) Fabric-retentive non-luminescent brachiopods as indicators of original $\delta13C$ and $\delta18O$ composition: a test. Journal of Sedimentary Petrology 60: 968-981

Rush P F, Chafetz H S (1991) Skeletal mineralogy of Devonian Stromatoporoids. Journal of Sedimentary Petrology 61: 364-369

Sandberg P A, Hudson J D (1983) Aragonite relic preservation in Jurassic calcite replaced bivalves. Sedimentology, 30: pp 879-892

Sandberg P A (1984) Recognition criteria for calcitized skeletal and non-skeletal aragonites. Palaeontographica Americana 54: 272-281

Sandberg P A (1985) Aragonite cements and their occurrence in ancient limestones. In: Schneidermann N, Harris P M (eds) Carbonate Cements. (SEPM Special Publication, vol 36). pp 33-57

Saller A H (1984) Petrological and geochemical constraints on the origin of subsurface dolomite, Enewetak Atoll: an example of dolomitization by normal seawater. Geology 12: 217-220

Sarg J F (1988) Carbonate sequence stratigraphy. In: Wilgus C K, Hastings B S, Kendall C G S C, Posamentier H W, Ross C A, van Wagoner J (eds) Sea-Level Changes: An Integrated Approach. (SEPM Special Publications, vol 42). pp 155-181

Schlager W (1981) The paradox of drowned reefs and carbonate platform. Bulletin of the Geological Society of America 92: 197-211

Schlager W, Austin J A, Comet P A, Droxler A W, Eberli G P, Freeman L R P, Fulthorpe C S, Harwood G M, Kuhn G, Lavoie D, Leckie R.-M, Melillo A J, Moore A, Mullins H T, Palmer A A, Ravenne C, Sager W W, Swart P K, Verbeek J W, Watkins D K, Williams C F (1985) Ocean Drilling Program; Rise and fall of carbonate platforms in the Bahamas. Nature 315: 632-633

Schlager W, Bourgeois F, Mackenzie G, Smit J (1988) Boreholes at Great Isaac and Site 626 and the history of the Florida Straits. In: Austin J A, Schlager W, Comet P A et al. (eds) Proceedings of the Ocean Drilling Program, Leg 101 (Scientific Report, vol 101). pp 425-437

Schlager W, Camber O (1986) Submarine slope angles, drowing unconformities and self-erosion of limestone escarpments. Geology 14: 762-765

Schlager W, Ginsburg R N (1981) Bahama carbonate platforms; the deep and the past. **Marine Geology** 44, pp 1-24

Schlager W, James N P, 1978) Low-magnesian calcite limestones forming at the deep-sea floor, Tongue of the Ocean, Bahamas. Journal of Sedimentology 25: 675-702

Schlager W, Reijmer J J G, Droxler A W (1994) Highstand shedding of carbonate platforms. Journal of Sedimentary Petrology B64: 270-281

Schlanger S O, Douglas R G (1974) The pelagic ooze-chalk-limestone transition and its implications for marine stratigraphy. In: Hsü K J, Jenkyns H C (eds) Pelagic Sediments on Land and under the Sea. (International Association of Sedimentologists Special Publications, vol 1). pp 117-148

Schmidt V (1965) Facies, diagenesis, and related reservoir properties in the Gigas Beds (Upper Jurassic), Northwestern Germany. In: Pray L C, Murray R C (eds) Dolimitization and Limestone Diagenesis. (SEPM Special Publications, vol 13). pp 124-168

Schmoker J W, Halley R B (1982) Carbonate porosity versus depth: a predictable relation for South Florida. AAPG Bulletin 66: 2561-2570

Schneidermann N, Sandberg P A (1971) Calcite-aragonite differentiation by selective staining and scanning electron microscopy. Annual Convention of Gulf Coast Association of Geological Societies Transactions 21: 349-352

Scholle P A (1979) Constituents, textures, cements and porosities of sandstones and associated rocks - a color illustrated guide. (AAPG Memoir, vol 28). AAPG, Tulsa/OK

Schuchert C (1935) Historical geology of the Antillean-Caribbean region or the lands bordering the Gulf of Mexico and the Caribbean Sea. Wiley and Sons, New York

Schuhmacher H (1977) Initial phases in reef development, studied at artificial reef types off Eilat, (Red Sea). Helgoländer Wissenschaftliche Meeresuntersuchungen 30: 400-411

Shackleton N J, Hall M A, Pate D (1995) Pliocene stable isotope stratigraphy of Site 846. In: Pisias N G, Mayer L A, Janecek T R et al. (eds) Proceedings of the Ocean Drilling Program, Leg 138. (Scientific Results, vol 138). pp 337-357

Sheridan R E (1971) Geotectonic evolution and subsidence of Bahama platform: Discussion. Geological Society of America Bulletin 82: 807-810

Sheridan R E (1974) Atlantic continental margin of North America. In: Burk C A, Drake C L (eds) Geology of Continental Margins. Springer, Berlin Heidelberg New York, pp 391-407

Sheridan R E (1976) Sedimentary basins of the Atlantic margin of North America. Tectonophysics 36: 113-132

Sheridan R E, Crosby J T, Bryan G M, Stoffa P L (1981) Stratigraphy and Structure of Southern Blake Plateau, Northern Florida Straits and Northern Bahama Platform from Multichannel Seismic Reflection Data. AAPG Bulletin 65: 2571-2593

Shinn E A, Halley R B, Hudson J H, Lidz B H (1977) Limestone compaction: an enigma. Geology 5: 21-24

Shinn E A, Robbin D M (1983) Mechanical and chemical compaction in fine-grained shallow-water limestones. Journal of Sedimentary Petrology 53: 595-618

Shinn E A, Steinen R P, Lidz B H, Swart P K (1989) Whitings, a sedimentologic dilemma. Journal of Sedimentary Petrology 59: 147-161

Shipboard Scientific Party (1997) Leg Synthesis: Sea-Level Changes and Fluid Flow on the Great Bahama Bank Slope. In: Eberli G P, Swart P K, Malone M J et al. (eds) Proceedings of the Ocean Drilling Program, Leg 166. (Initial Reports, vol 166). pp 13-22

Silver B A, Todd R G (1969) Permian cyclic strata, northern Midland and Delaware Basins, west Texas and southeastern New Mexico. AAPG Bulletin 53: 2223-2251

Simms M (1984) Dolomitization by groundwater flow systems in carbonate platforms. Transactions Gulf Coast Association of Geological Societies 24: 411-420

Sorby H C (1879) The structure and origin of limestones. Anniversary address of the president. Quarterly Journal of the Geological Society of London 35: 56-95

Steinen R P, 1978) On the diagenesis of lime mud: Scanning electron microscopic observations of surface material from Barbados, W.I.. Journal of Sedimentary Petrology 48: 1139-1147

Steinen R P, 1982) SEM observations on the replacement of Bahaman aragonitic mud by calcite. Geology 10: 471-475

Steinen R P, Swart P K, Shinn E A, Lidz B H (1988) Bahamian lime mud: the algae didn't do it. Geological Society of America 1988 centennial celebration, Abstracts with Programs 20/7: A209

Stevens J (1726) The General History of the Vast Continent and Islands of America

Stoakes F A (1980) Nature and control of shale basin fill and its effect on reef growth and termination: Upper Devonian Duvernay and Ireton Formations of Alberta, Canada. Bulletin of Canadian Petroleum Geology 28: 345-410

Stockman K W, Ginsburg R N, Shinn E A (1967) The production of lime mud by algae in south Florida. Journal of Sedimentary Petrology 37: 633-648

Suess E (1888) Das Antlitz der Erde. (vol 2). F Tempsky, Vienna

Suess E (1908) Das Antlitz der Erde. (vol 1). F Tempsky, Vienna

Supko P R (1963) A quantitative X-ray Diffraction Method for Mineralogical Analysis of Carbonate Sediments from Tongue of the Ocean. Dissertation, University of Miami

Swart P K, Elderfield H, Beets K (in press-a) The $^{87}Sr/^{86}Sr$ ratios of carbonates, phosphorites and fluids collected during the Bahama Drilling Project cores Clino and Unda: Implications for dating and diagenesis. In: Ginsburg R N (ed) Ground Truthing Seismic Stratigraphy of a Prograding Carbonate Platform Margin: Neogene, Great Bahama Bank. (SEPM Contributions in Sedimentology)

Swart P K, Elderfield H, Ostlund G (in press-b) The geochemistry of pore fluids from bore holes in the Great Bahama Bank. In: Ginsburg R N (ed) Ground Truthing Seismic Stratigraphy of a

Prograding Carbonate Platform Margin: Neogene, Great Bahama Bank. (SEPM Contributions in Sedimentology)

Swart P K, Guzikowski M (1988) Interstitial water chemistry and diagenesis of periplatform sediments from the Bahamas, ODP Leg 101. In: Austin J A, Schlager W, Comet P A et al. (eds) Proceedings of the Ocean Drilling Program, Leg 101. (Scientific Results, vol 101). pp 363-380

Swinchatt J P (1969) Algal Boring: A Possible Depth Indicator in Carbonate Rocks and Sediments. Geological Society of America Bulletin 80: 1391-1396

Talwani M, Worzel J L, Ewing M (1960) Gravity anomalies and structure of the Bahamas. 2nd Caribbean Geol. Conf. Trans., University of Puerto Rico: 156-160

Tarutani T, Clayton R N, Mayede T K (1969) The effect of polymorphism and magnesium substitution on oxygen isotope fractionation between calcium carbonate and water. Geochimica et Cosmochimica Acta 33: 987-996

Tennant C B, Berger R W (1957) X-Ray determination of dolomite-calcite ratio at a carbonate rock. American Mineralogist 42: 23-29

Terzaghi R D (1940) Compaction of lime mud as a cause of secondary structure. Journal of Sedimentary Petrology 10: 78-90

Traverse A, Ginsburg R N (1966) Palynology of the surface sediments of Great Bahama Bank, as related to water movement and sedimentation. Marine Geology 4: 417-459

Uchupi E, Milliman J D, Luyendyk B P, Brown C O, Emery K O (1971) Structure and origin of southeastern Bahamas. AAPG Bulletin 55: 687-704

Vail P R, Audemard F, Bowman S A, Eisner N, Perez-Cruz C (1991) The Stratigraphic Signatures of Tectonics, Eustacy and Sedimentology - an Overview. In: Einsele G, Ricken W, Seilacher A (eds) Cycles and Events in Stratigraphy. Springer, Berlin Heidelberg New York, pp 617-659

Van der Plas L, Tobi A C (1965) A chart for judging the reliability of point-counting results. American Journal of Science 263: 87-90

Vaughan T W (1910) Preliminary Remarks on the Geology of the Bahamas, with Special Reference to the Origin of the Bahaman and Floridian Oölites. Pap. Tortugas Lab. 182: 47-54

Veizer J (1977) Diagenesis of pre-Quaternary carbonates as indicated by tracer studies. Journal of Sedimentary Petrology 47: 565-581

Veizer J (1992) Depositional and diagenetic history of limestones: stable and radiogenic isotopes. In: Clauer N, Chandhuri S (eds) Isotopic signatures and sedimentary records. (Lecture Notes in Earth Science, vol 43). Springer, Berlin Heidelberg New York, pp 13-48

Vine P J, Bailey-Brock J H (1984) Taxonomy and ecology of coral reef tube worms (Serpulidae, Spirorbidae) in the Sudanese Red Sea. Zoological Journal of the Linnean Society 80: 135-156

Wall D, Dale B, Harada K (1973) Descriptions of new fossil dinoflagellates from the Late Quaternary of the Black Sea. Micropaleontology 19: 18–31

Wall D, Dale B, Lohmann G P, Smith W K (1977) The environmental and climatic distribution of dinoflagellate cysts in modern marine sediments from regions in the north and south Atlantic Oceans and adjacent seas. Marine Micropaleontology 2: 121-200

Walther M (1982) A Contribution to the Origin of Limestone-Shale Sequences. In: Einsele G, Seilacher A (eds.): Cyclic and Event Stratification. Springer, Berlin Heidelberg New York, pp 113-120

Webb P-N, Harwood D M (1991) Late Cenozoic glacial history of the Ross embayment, Antarctica. Quaternary Science Reviews 10: 215-223

Weber J N, Smith F G (1961) Rapid determination of calcite-dolomite ratios in sedimentary rocks. Journal of Sedimentary Petrology 31: 130-131

Weller J M (1959) Compaction of sediments. AAPG Bulletin 43: 273-310

Westphal H, Munnecke A (1997) Mechanical compaction versus early cementation in fine-grained limestones: differentiation by the preservation of organic microfossils. Sedimentary Geology 112: 33-42

Whitaker F F, Smart P L (1990) Active circulation of saline ground waters in carbonate platforms: evidence from the Great Bahama Bank. Geology 18: 200-203

Whitaker F, Smart P, Hague Y, Waltham D, Bosence D (1997) Coupled two-dimensional diagenetic and sedimentological modeling of carbonate platform evolution. Geology 25: 175-178

Whitaker F F, Smart P L, Vahrenkamp V C, Nicholson H, Wogelius R A (1994) Dolomitization by normal sea-water? Field evidence from the Bahamas. In: Purser B, Tucker M, Zenger D (eds) Dolomites - A volume in honour of Dolomieu. (International Association of Sedimentologists Special Publication, vol 21). pp 111-132

Wilber R J, Milliman J D, Halley R B (1990) Accumulation of bank-top sediment on the western slope of Great Bahama Bank: Rapid progradation of a carbonate megabank. Geology 18: 970-974

Williams S C (1985) Stratigraphy, Facies Evolution and Diagenesis of Late Cenozoic Limestones and Dolomites, Little Bahama Bank, Bahamas. Dissertation, University of Miami

Wilson J L (1975) Carbonate facies in geologic history. Springer, Berlin Heidelberg New York

Wilson P A, Roberts H H (1992) Carbonate-periplatform sedimentation by density flows: a mechanism for rapid off-bank and vertical transport of shallow-water fines. Geology 20: 713-716

Wilson P A, Roberts H H (1995) Density cascading: off-shelf sediment transport, evidence and implications, Bahama Banks. Journal of Sedimentary Research A65: 45-56

Wolf K H, Conolly J R (1965) Petrogenesis and paleoenvironment of limestone lenses in Upper Devonian Red Beds of New South Wales. Palaeogeography-Palaeoclimatology-Palaeoecology 1: 69-111

Wray J L (1977) Calcareous algae. (Developments in Paleontology and Stratigraphy, vol 4). Elsevier, Amsterdam

Zeff M L, Perkins R D (1979) Microbial alteration of Bahamian deep sea carbonates. Journal of Sedimentology 26: 175-201

Appendix 1

Photographical Plates

PLATE 1

Components of the selected Upper and Lower Pliocene intervals from CLINO.—

(A) *Halimeda* plate. The irregular tubes are clearly visible as micritic linings. (Upper Pliocene; 256.18 mbmp)

(B) Red algal encrustation (Melobesioida). Regular cellular structure is largely obscured by diagenesis. (Upper Pliocene; 256.79 mbmp)

(C) Coralline algae. Characteristic fine-cellular microstructure appears well preserved on this level of magnification. (Upper Pliocene; 220.26 mbmp)

(D) SEM micrograph of single coccolith (possibly *Umbilicosphaera sp.*). Coccoliths are extremely rare in the Pliocene periplatform sediments. (Lower Pliocene; 476.55 mbmp)

(E) Neomorphosed, originally aragonitic gastropod. The skeletal structure of the shell is entirely lost. Gastropods found in the Pliocene samples examined usually are neomorphous. (Upper Pliocene; 256.59 mbmp).

(F) Bivalve shell showing poorly preserved internal structure (layers). (Upper Pliocene; 256.18 mbmp).

(G) Encrusting bryozoan (possibly *Schizoporella*) with geopetally filled chambers. (Upper Pliocene; 256.49 mbmp)

(H) Bryozoan *Adeona* exhibiting the typical bisymmetrical shape. Above *Adeona*, a bivalve fragment shows severe bioerosion. (Upper Pliocene; 220.80 mbmp)

Plate 1   173

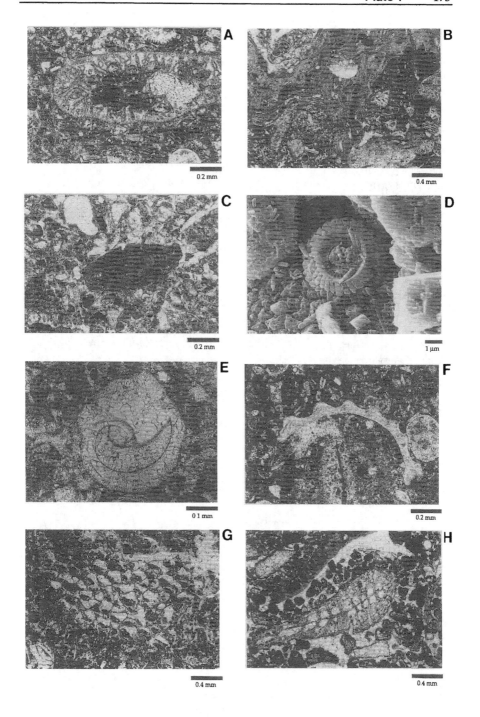

PLATE 2

Components of the selected Upper and Lower Pliocene intervals from CLINO; continued.— Representative palynomorphs (various magnifications): (A)-(E) are non-dinoflagellate palynomorphs, (F)-(R) are dinoflagellate cysts, and (S) is a view of the palynofacies at lower magnification. Micrographs by M. J. Head.

(A) Invertebrate remains, probable scolecodont. Length 51 μm (477.77 mbmp)

(B) Trochospiral micro-foraminiferal lining. Maximum diameter 128 μm (Lower Pliocene; 476.40 mbmp)

(C) Trilete fern spore. Diameter 59 μm. (Upper Pliocene; 255.88 mbmp)

(D) Bisaccate pollen of the genus Pinus. Length 49 μm. (220.48 mbmp)

(E) Tetracolporate angiosperm pollen. Polar diameter 27 μm. (Upper Pliocene; 256.79 mbmp)

(F) *Dapsilidinium pseudocolligerum* (Stover, 1977). Central body maximum diameter 42 μm. (Upper Pliocene; 232.71 mbmp)

(G) *Hystrichokolpoma rigaudiae* (Deflandre and Cookson, 1955). Central body width 50 μm. (Lower Pliocene; 494.23 mbmp)

(H) *Impagidinium paradoxum* (Wall, 1967). Central body length 32μm. (Lower Pliocene; 476.40 mbmp)

(I) *Kallosphaeridium sp.* showing apical archeopyle suture. Maximum diameter 32 μm. (Lower Pliocene; 477.77 mbmp)

(J) *Lingulodinium machaerophorum* (Deflandre & Cookson, 1955). Central body maximum diameter 46μm. (Lower Pliocene; 476.40 mbmp)

(K) *Melitasphaeridium choanophorum* (Deflandre & Cookson, 1955). Central body maximum diameter 31 μm. (Upper Pliocene; 234.12 mbmp)

(L) *Nematosphaeropsis rigida* (Wrenn, 1988). Central body length 38 μm. (Upper Pliocene; 255.88 mbmp)

(M) *Operculodinium israelianum* (Rossignol, 1962). Central body maximum diameter 61 μm. (Lower Pliocene; 494.23 mbmp)

(N) *Operculodinium longispinigerum* (Matsuoka, 1983). Central body length, 34 μm. (Lower Pliocene; 477.77 mbmp)

(O) *Spiniferites sp.* Central body length 48 μm. (Lower Pliocene; 476.40 mbmp)

(P) *Selenopemphix quanta* (Bradford, 1975). Central body maximum diameter 47 μm. (Upper Pliocene; 234.12 mbmp)

(Q) *Polysphaeridium zoharyi* (Rossignol, 1962). Central body maximum diameter 58 μm. (Lower Pliocene; 476.40 mbmp)

(R) *Tuberculodinium vancampoae* (Rossignol, 1962). Maximum diameter 105 μm. (Lower Pliocene; 476.40 mbmp)

(S) Typical palynofacies, after brief ultrasonification and sieving at 10 micrometers, showing abundant fragmented micro-foraminiferal linings as darker subspherical objects, and lighter membranous debris of algal and possibly also foraminiferal origin. Note the virtual absence of terrigenous material. Field of view length, 330 μm. (Upper Pliocene; 226.01 mbmp)

Plate 2   175

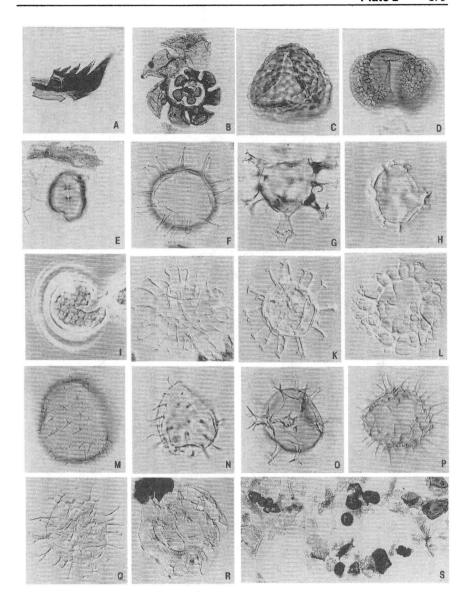

PLATE 3

Components of the selected Upper and Lower Pliocene intervals from CLINO; continued.—

(A) Small miliolid foraminifer, common in very shallow, warm and slightly restricted marine waters. The test of this internally cemented specimen is dissolved leaving a delicate mold. (Upper Pliocene; 259.23 mbmp)

(B) Large miliolid foraminifer (*Pyrgo*). The test is preserved and is characterized by a dark color in thin section, typical for miliolid foraminifers. (Upper Pliocene; 256.59 mbmp)

(C) Textulariid foraminifer (*Bigenerina*) showing the typical agglutinated wall and bisymmetrical shape. (Upper Pliocene; 255.88 mbmp)

(D) Rotaliid foraminifer (possibly *Rotalia*). The clear test is characteristic for rotaliid foraminifers. (Upper Pliocene; 261.75 mbmp)

(E) Small *Rosalina* test. In the juvenile chambers, the inner organic layer is preserved, appearing dark red in thin section. (Upper Pliocene; 246.30 mbmp)

(F) *Amphistegina*, typical swollen, massive test with small chambers. (Upper Pliocene; 256.49 mbmp)

(G) Planktic foraminifer (probably *Globigerinoides*). (Upper Pliocene; 219.94 mbmp)

(H) Planktic foraminifer *Orbulina*. The internal juvenile chambers are clearly discernible. In the lower left corner, a small *Rosalina* test is seen. (Lower Pliocene; 461.19 mbmp)

Plate 3   177

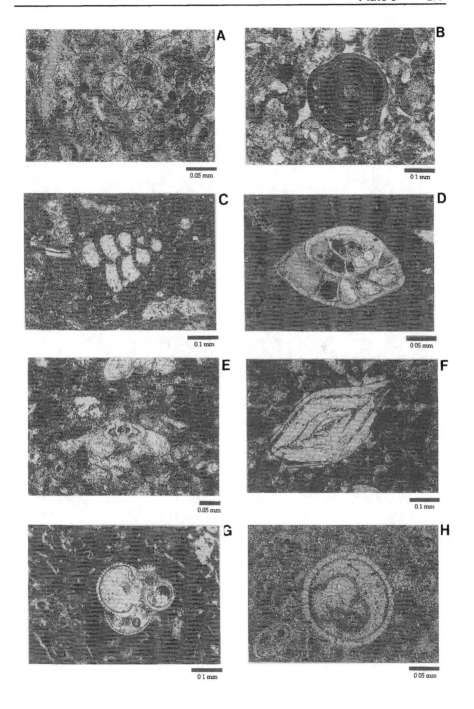

PLATE 4

Components of the selected Upper and Lower Pliocene intervals from CLINO; continued.—

(A) Ostracod test with homogeneous prismatic wall structure. At the left end, the two valves show a slight overlap that is common in ostracods. (Upper Pliocene; 219.94 mbmp)

(B) Longitudinal section through echinoderm spine. At the right end, the original striate skeletal structure is visible. The spine is enclosed by clear syntaxial cement that under crossed nicols shows unit extinction. (Upper Pliocene; 221.34 mbmp)

(C) Borings in rotaliid test of endolithic boring organisms. (Lower Pliocene; 508.25 mbmp)

(D) SEM micrograph of a peloid with traces of soft substrate burrowers, indicating that the peloid is a fecal pellet. (Upper Pliocene; 262.18 mbmp)

(E) Non-determinable grains that are strongly altered (recrystallized) by diagenesis. The altered grains could represent originally aragonitic skeletal material (scleractinians). (Lower Pliocene; 503.99 mbmp)

(F) Cortoid characterized by a micritic rim. The internal structures of the skeletal grain are still present as relic structures). (Upper Pliocene; 256.82 mbmp)

(G) Intraclast of partly consolidated sediment, consisting of a dominantly micritic material. Incorporated fine grains are indented rather than cut by the boundaries of the intraclasts. This indicates that the material was still soft at the time of reworking. (Upper Pliocene; 256.95 mbmp)

(H) SEM micrograph of aragonite needles that constitute the matrix of many fine-grained samples. Sample is a broken surface and is not etched. (Upper Pliocene; 256.79 mbmp)

Plate 4   179

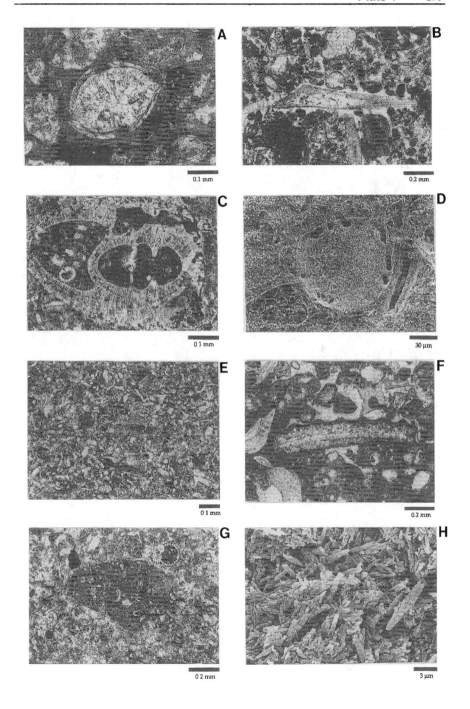

PLATE 5

Microfacies from the Upper Pliocene interval. All micrographs show same magnification.—

(A) Microfacies 1: cortoid grainstones. Cortoids are a dominant constituent. Sedimentary matrix is rare to absent. (Upper Pliocene; 221.34 mbmp)

(B) Microfacies 2: cortoid packstones. Similar to microfacies 1, but smaller-grained and contains fine-grained sedimentary matrix. (Upper Pliocene; 256.95 mbmp)

(C) Microfacies 3, *Halimeda*-rich biomicrites. The large *Halimeda* plates (**h**) appear light-colored as they are slightly neomorphosed, whereas the sedimentary matrix appears dark. (Upper Pliocene; 256.18 mbmp)

(D) Microfacies 4: nodule-rich biomicrites. Red algal nodules (left part of micrograph) make up 21% of this microfacies. An *Amphistegina* test is seen below the middle of micrograph. (Upper Pliocene; 256.79 mbmp)

(E) Microfacies 5: miliolid packstones. Up to 210 miliolids per cm^2 (white arrows) are found in thin sections of microfacies 5. (Upper Pliocene; 261.51 mbmp)

(F) Microfacies 6: mixed pack- to wackestones. Peloids and skeletal debris are equally abundant. (Upper Pliocene; 262.59 mbmp)

(G) Microfacies 7: peloid packstones. Peloids are the dominant constituent. Usually the peloid-dominated samples are strongly compacted as is indicated by a broken foraminifer test (white arrow). Peloids are strongly deformed, sometimes leading to the formation of a "pseudomatrix". (Upper Pliocene; 261.82 mbmp)

(H) Microfacies 8: biodetrital packstones. Peloids are rare, and biodetrital grains (e.g. rotaliid foraminifer in the center, echinoderm fragment to the right) are predominant. (Upper Pliocene; 259.23 mbmp)

Plate 5   181

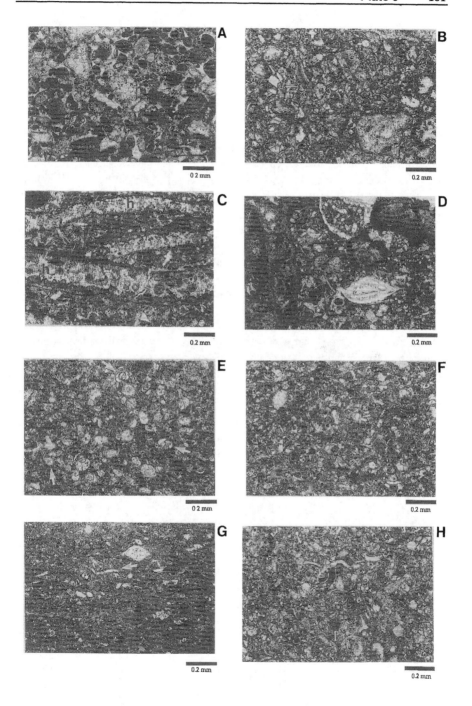

PLATE 6

Microfacies from the Upper Pliocene interval (continued). All micrographs show same magnification.—

(A) Microfacies 9: biodetrital wackestones. Non-determinable debris is abundant, but also discernible components are present, e.g. small miliolid and textulariid foraminifers. (Upper Pliocene; 255.27 mbmp)
(B) Microfacies 10: mudstones. In the fine-grained material, non-determinable grains, peloids, and foraminifers (center of micrograph) occur. Other constituents are rare. (Upper Pliocene; 226.31 mbmp)

Microfacies from the Lower Pliocene interval. All micrographs show same magnification.—

(C) Microfacies 11: bioclast-peloid packstones. Diagenetically strongly altered biodetritus is a dominant constituent, but also peloids occur in considerable amounts. They often are found to form clusters. (Lower Pliocene; 490.19 mbmp)
(D) Microfacies 12: bioclast packstones. Strongly altered small-sized biodetrital grains constitute most of this microfacies. (Lower Pliocene; 503.99 mbmp)
(E) Microfacies 13: bioclast wackestones. Compaction varies in this microfacies; the sample shown is characterized by apparent lamination induced by compaction. (Lower Pliocene; 453.24 mbmp)
(F) Microfacies 14: globigerinid packstone. The strongly open marine signature of this microfacies is seen in the large amount of planktic foraminifers. Typical platform derived material is largely absent. (Lower Pliocene; 509.47 mbmp)

Plate 6 183

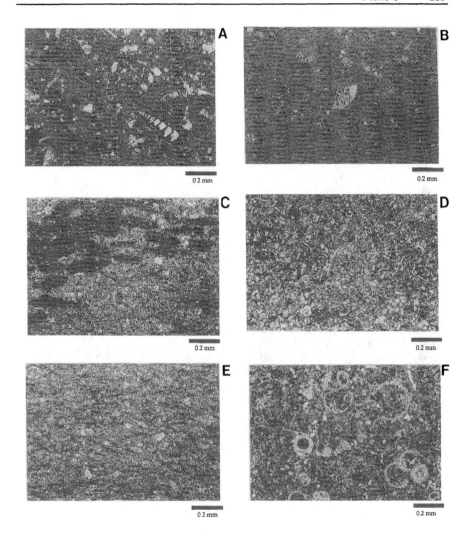

PLATE 7

Matrix constituents - SEM observations.—

(A) Micritic grains and larger crystals in an uncemented sample from the Lower Pliocene. m = micrite, d = dolomite, c = larger calcite crystal. (Lower Pliocene; 476.15 mbmp)

(B) Tight mosaic of microspar cements enclosing aragonite needles. (Upper Pliocene; 262.18 mbmp)

(C) Partly lithified sample. Some of the microspar cements exhibit a subhedral shape. Aragonite needles, being enclosed in the cements are preserved. (Upper Pliocene; 217.04 mbmp)

(D) Calcite grain (e = echinoderm fragment?) with cement overgrowth (c). Aragonite needles are enclosed in the syntaxial cement. (Upper Pliocene; 217.04 mbmp)

(E) LMC shell fragment bordered sharply by microspar crystals. (Upper Pliocene; 234.12 mbmp)

(F) Pitted microspar. No aragonite needles are observed in this sample. (Lower Pliocene; 497.74 mbmp)

(G) Geopetally filled pteropod shell with transistion from sedimentary filling that is cemented by microspar, and sparitic filling with smooth surface. Some crystals belong to both, the (microsparitic) sedimentary layer (m) and the sparitic filling (s). (Upper Pliocene; 242.77 mbmp)

(H) Planktonic foraminifer with syntaxial internal cements. Microspar cements at the outside of the shell locally also show a syntaxial direction. (Lower Pliocene; 494.23 mbmp)

Plate 7   185

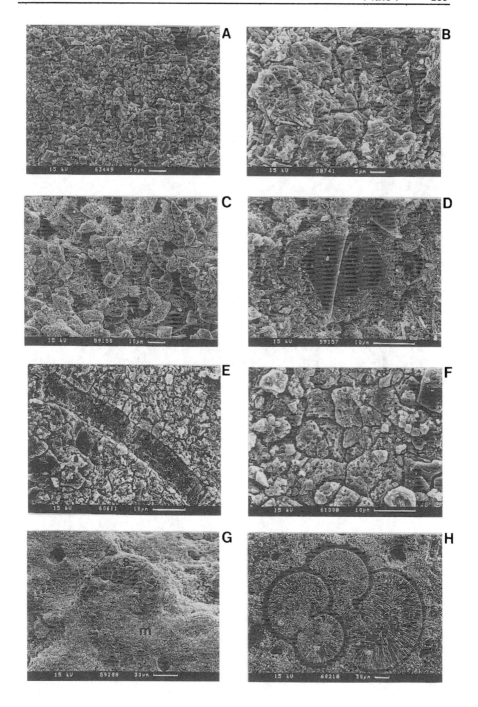

PLATE 8

Diagenetic calcite, dolomite, celestite, and phosphate.—

(A) Acicular cement in shell. Blocky cements constitutes a second generation of cement, enclosing the early acicular cements. The original shell is dissolved. (Upper Pliocene; 262.18 mbmp)

(B) Dolomite rhomboids in the matrix of a Lower Pliocene sample. (Lower Pliocene; 490.19 mbmp)

(C) Benthic foraminifer with dolomite cement infilling the pores of the wall. The structure (layers) of the test is well preserved. (Upper Pliocene; 257.35 mbmp)

(D) Planktonic foraminifer with dolomite infilling the pores of the wall and covering the inside of the test. The outside (matrix) is composed of fine-grained dolomite. The internal cement is microsparitic. (Upper Pliocene; 257.35 mbmp)

(E) Bioclast with endolithic borings. Borings are infilled by dolomite cement. (Lower Pliocene; 495.30 mbmp)

(F) Celestite matrix with moldic pore, indicating that the celestite has precipitated as a primary cement. (Lower Pliocene; 490.19 mbmp)

(G) *Halimeda* plate preserved as celestite. (Upper Pliocene; 256.03 mbmp)

(H) Phosphatized matrix. Bioclastic grains are preserved, being bound by the cryptocrystalline phosphatic matrix. (Lower Pliocene; 509.17 mbmp)

Plate 8 187

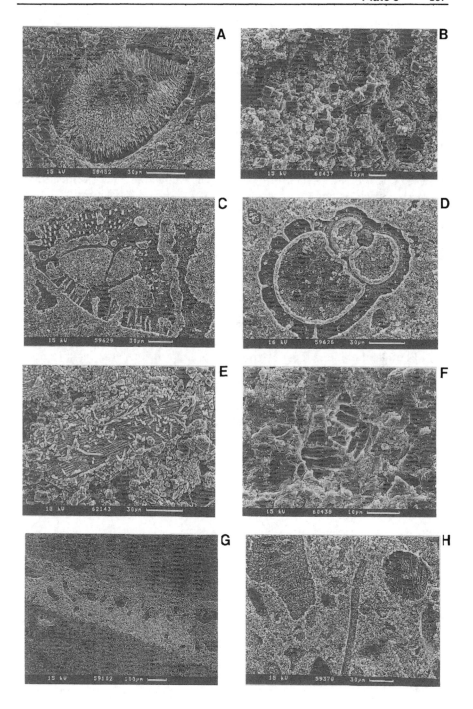

PLATE 9

Diagenesis of fossils.—

(A) Dolomitized red alga. A distinction between primary and secondary layers is difficult as both appear dolomitized. (Upper Pliocene; 220.83 mbmp)

(B) Dolomitized red alga. Primary layers of the wall are dolomitized, whereas the secondary layers appear as gap. (Upper Pliocene; 261.98 mbmp)

(C) Detail of *Halimeda* plate. Aragonite needles around tubes are cemented by microspar. Space between tubes is void. (Upper Pliocene; 256.18 mbmp)

(D) Detail of *Halimeda* plate. Specimen is tightly cemented by microspar crystals that enclose the aragonite needles of the calcareous alga. Such inclusions could account for the pseudopleochroism as seen in the light microscope. (Upper Pliocene; 256.03 mbmp)

(E) Echinoderm spine. Primary pores in the spine are still open. (Upper Pliocene; 261.98 mbmp)

(F) Detail of echinoderm spine (E). Small inclusions are conspicuous in the skeletal crystal. The spine is preserved as LMC. The inclusions are interpreted as microdolomite as result of unmixing. (Upper Pliocene; 261.98 momp)

(G) Miliolid foraminifer. Test of miliolid is hardly visible. It is interpreted to have been dissolved after precipitation of internal cements. Afterwards the mold was closed by the growth of the internal cement. On the outside of the foraminifer, microspar is characterized by enclosed aragonite needles, on the inside, microspar to sparite crystals without inclusions occur. (Upper Pliocene; 225.70 mbmp)

(H) Rotaliid foraminifer with well preserved inner organic layer. (Upper Pliocene; 261.82 mbmp)

Plate 9   189

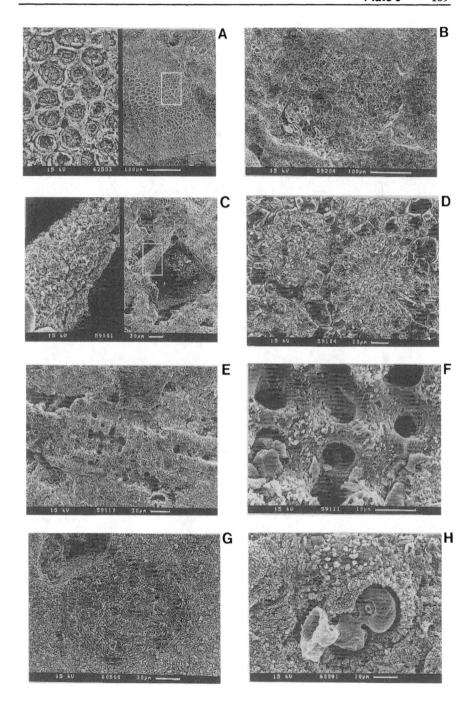

PLATE 10

Thin-walled organic microfossils as compaction indicators.—

(A) Dinoflagellate cyst *Polysphaeridium zoharyi* in spherical preservation. The cysts is internally hollow, implying that early cementation of the matrix prevented deformation of the organic-walled cyst. (Lower Pliocene; 497.89 mbmp)
(B) Dinoflagellate cyst in spherical preservation, showing internal cementation. (Upper Pliocene; 226.01 mbmp)
(C) Deformed dinoflagellate cyst in uncemented Upper Pliocene sample that is characterized by high amounts of aragonite needles composing the matrix. (Upper Pliocene; 253.14 mbmp)
(D) Deformed dinoflagellate cyst from an uncemented Lower Pliocene sample. (Lower Pliocene; 484.02 mbmp)

Plate 10 191

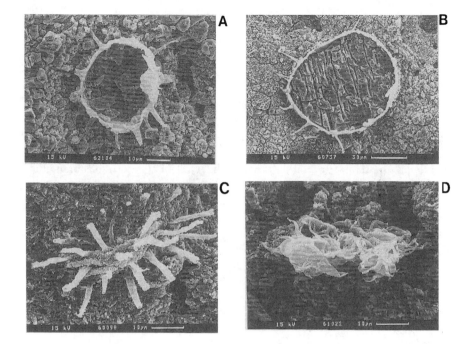

PLATE   11

Light microscopic observations on diagenesis.—

(A) Peloidal cement inside planktonic foraminifer. Triangular test in the lower left is foraminifera *Reusella*. (Upper Pliocene; 256.03 mbmp)

(B) Dogtooth shaped internal cement in *Orbulina*. (Lower Pliocene; 509.17 mbmp)

(C) Small dog-tooth fringe and sparry secondary cement in miliolid foraminifer. Crossed nicols. (455.98 mbmp)

(D) ?echinoderm fragment with syntaxial cement. Syntaxial cement seems to have grown faster and thus reached a larger size than adjacent cement crystals. (Upper Pliocene; 257.35 mbmp)

(E) *Halimeda* fragment (detail). Acicular cements are observed. Most of *Halimeda* is cemented by sparite. As the sparry cement appears clear, most of the skeletal aragonite needles probably are dissolved. Remains of the tubes are still discernible. (Upper Pliocene; 256.79 mbmp)

(F) Planktic foraminifer with internal pyrite framboids. (Upper Pliocene; 253.44 mbmp)

(G) *Halimeda* fragment with cemented tubes. The space between the tubes is void. (Upper Pliocene; 256.18 mbmp)

(H) Planktonic foraminifer that is collapsed due to compaction. Note that in none of the foraminifers (neither the large collapsed one nor the small ones) present cement crystals are observed. (Upper Pliocene; 261.82 mbmp)

Plate 11   193

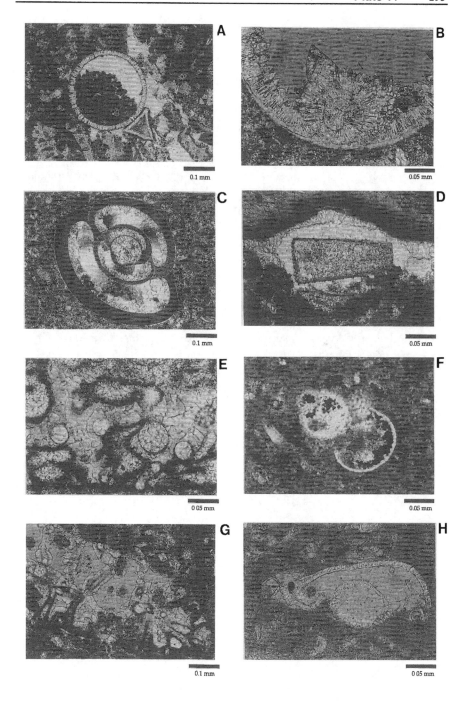

PLATE 12

Ultrafacies. All micrographs show same magnification.—

(A) Ultrafacies 1: Uncemented aragonite needle mesh with some crystallites and dolomite rhomboids (1μm in length). (Upper Pliocene; 219.94 mbmp)

(B) Ultrafacies 2: Partly cemented aragonite needle mesh. Microspar precipitates enclose aragonite needles. (Upper Pliocene; 217.04 mbmp)

(C) Ultrafacies 3: Tight mosaic of microspar cements that engulf aragonite needles. (Upper Pliocene; 217.17 mbmp)

(D) Ultrafacies 4: Tight mosaic of pitted microspar crystals. (Lower Pliocene; 497.74 mbmp)

(E) Ultrafacies 5: Micritic, uncemented sample with some larger crystallites. (Lower Pliocene; 479.82 mbmp)

(F) Ultrafacies 6: Transitional between ultrafacies 4 and 5, this sample shows local cementation. (Lower Pliocene; 464.49 mbmp)

(G) Ultrafacies 7: Tight sparitic cementation without aragonite needle inclusions. (Upper Pliocene; 220.83 mbmp)

(H) Ultrafacies 8: Dense, fine-grained dolomite composes the matrix. (Upper Pliocene; 257.24 mbmp)

Plate 12   195

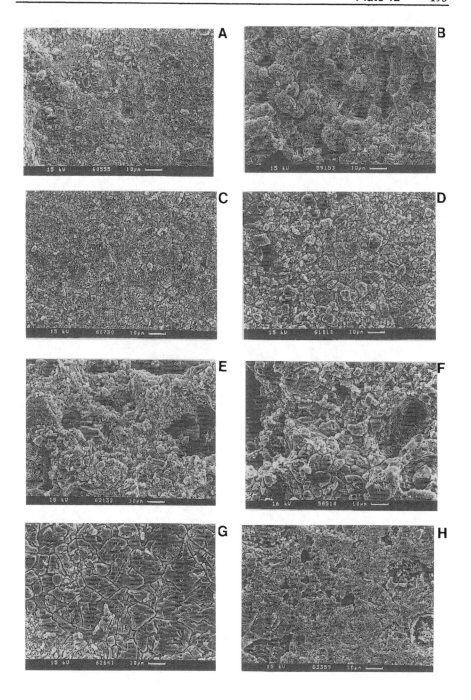

Palynomorph Assemblages

determined by M. J. Head

Appendix 2

Taxa\sample [mbmp]	Plate 2	220.48	226.01	232.71	234.12	255.88	256.79	261.67	262.18	474.40	476.61	477.04	477.77	494.23	494.39
Achomosphaera/Spiniferites spp. (total)	O	57	159	173	151	136	42	137	161	86	139	131	175	42	109
Spiniferites mirabilis		•		•	•	•		•	•	•	•	•		•	
Spiniferites (ramified process bases)		•		•	•	•		•	•	•			•	•	•
Ataxiodinium zevenboomii			(+)?		+	+	+	+	+				1		
Bitectatodinium raedwaldii				+		+		3	+	+	+	+			
Bitectatodinium cf. raedwaldii						+	+								
Bitectatodinium tepikiense										1					
Capisocysta lata		1	29	15	35	1	2	11	13	5	2	8	3	+	+
Dapsilidinium pseudocolligerum	F	1	10	4	6	3	1	3	1		+		1	2	2
Protoperidinioid cyst sp. A						+									
Hystrichokolpoma rigaudiae	G									+	+		+	1	+
Impagidinium paradoxum	H						3	+		+					
Impagidinium plicatum													+		
Kallosphaeridium sp.	I	•		•	•	•			•	•	•	•	•		
Lejeunecysta marieae							+								
Lejeunecysta sp. cf. L. marieae		2													
Lingulodinium machaerophorum	J	7	25	33	28	77	2	56	42	13	15	18	5	33	16
Melitasphaeridium choanophorum	K		1	6	17	10									
Nematosphaeropsis rigida	L					4									
Operculodinium spp. including israelianum	M	3	16	11	9	9	3	33	29	24	42	32	24	9	12
Operculodinium longispinigerum	N									+	1	1	+		1
Operculodinium bahamaense Head n. sp.		+	5	1	+	3		3	+	+	+	+			+
Operculodinium psilatum				+										+	+
Operculodinium janduchenei													+		
Polysphaeridium zoharyi	Q	27				4	42	+	+	116	49	47	33	160	104
Round brown cysts		2	2	6	3	+	1	3	1	4		12	6	3	4
Selenopemphix quanta	P		+	1	+	1	1	1	2		+	+	+	+	2
Selenopemphix nephroides			+		+	+						+			+
Tuberculodinium vancampoae	R		3		1					1	2	1	2	+	+
Operculodinium? megagranum Head n. sp.						2	+								
TOTAL IN-SITU DINOFLAGELLATES		100	250	250	250	250	97	250	249	250	250	250	250	250	250
Spike = quantity of marker grains		1933	605	117	176	241	3414	278	95	173	35	90	94	325	27
Marine algae incertae sedis															
Cyclopsiella sp.		1				1									
Nanobarbophora walldalei		3	19	9	44	1		29	3	14	51	52	32	5	18
Small spiny acritarchs		16	23	51	63	68	25	108	43	93	112	131	158	11	43
Acritarch sp. 1		•	•				•		•						
Incertae sedis sp. A													+	1	
Incertae sedis sp. B										+					
In-situ palynomorphs															
In-situ dinoflagellates															
Marine algae incertae sedis															
Bisaccate pollen	D	88	3	2	2	6	27	4				1	1	4	1
Angiosperm pollen	E	9	+			3	6			1		1	1		
Fern and bryophyte spores	C			+		1	+	1			1				
Fungal spores and hyphae						1	2								
Foraminiferal linings (6 or more chambers)	B	6	10	6	9	12		15	9	34	31	19	31	22	11
Scolecodonts	A	1					1			4		2	1		1
Copepod fragments				+		+								1	
Copepod egg fragments?					+					3					

Asterisk indicates "present but not counted"
Cross indicates "present in low amounts"

Lecture Notes in Earth Sciences

Vol. 37: A. Armanini, G. Di Silvio (Eds.), Fluvial Hydraulics of Mountain Regions. X, 468 pages. 1991.

Vol. 38: W. Smykatz-Kloss, S. St. J. Warne, Thermal Analysis in the Geosciences. XII, 379 pages. 1991.

Vol. 39: S.-E. Hjelt, Pragmatic Inversion of Geophysical Data. IX, 262 pages. 1992.

Vol. 40: S. W. Petters, Regional Geology of Africa. XXIII, 722 pages. 1991.

Vol. 41: R. Pflug, J. W. Harbaugh (Eds.), Computer Graphics in Geology. XVII, 298 pages. 1992.

Vol. 42: A. Cendrero, G. Lüttig, F. Chr. Wolff (Eds.), Planning the Use of the Earth's Surface. IX, 556 pages. 1992.

Vol. 43: N. Clauer, S. Chaudhuri (Eds.), Isotopic Signatures and Sedimentary Records. VIII, 529 pages. 1992.

Vol. 44: D. A. Edwards, Turbidity Currents: Dynamics, Deposits and Reversals. XIII, 175 pages. 1993.

Vol. 45: A. G. Herrmann, B. Knipping, Waste Disposal and Evaporites. XII, 193 pages. 1993.

Vol. 46: G. Galli, Temporal and Spatial Patterns in Carbonate Platforms. IX, 325 pages. 1993.

Vol. 47: R. L. Littke, Deposition, Diagenesis and Weathering of Organic Matter-Rich Sediments. IX, 216 pages. 1993.

Vol. 48: B. R. Roberts, Water Management in Desert Environments. XVII, 337 pages. 1993.

Vol. 49: J. F. W. Negendank, B. Zolitschka (Eds.), Paleolimnology of European Maar Lakes. IX, 513 pages. 1993.

Vol. 50: R. Rummel, F. Sansò (Eds.), Satellite Altimetry in Geodesy and Oceanography. XII, 479 pages. 1993.

Vol. 51: W. Ricken, Sedimentation as a Three-Component System. XII, 211 pages. 1993.

Vol. 52: P. Ergenzinger, K.-H. Schmidt (Eds.), Dynamics and Geomorphology of Mountain Rivers. VIII, 326 pages. 1994.

Vol. 53: F. Scherbaum, Basic Concepts in Digital Signal Processing for Seismologists. X, 158 pages. 1994.

Vol. 54: J. J. P. Zijlstra, The Sedimentology of Chalk. IX, 194 pages. 1995.

Vol. 55: J. A. Scales, Theory of Seismic Imaging. XV, 291 pages. 1995.

Vol. 56: D. Müller, D. I. Groves, Potassic Igneous Rocks and Associated Gold-Copper Mineralization. 2nd updated and enlarged Edition. XIII, 238 pages. 1997.

Vol. 57: E. Lallier-Vergès, N.-P. Tribovillard, P. Bertrand (Eds.), Organic Matter Accumulation. VIII, 187 pages. 1995.

Vol. 58: G. Sarwar, G. M. Friedman, Post-Devonian Sediment Cover over New York State. VIII, 113 pages. 1995.

Vol. 59: A. C. Kibblewhite, C. Y. Wu, Wave Interactions As a Seismo-acoustic Source. XIX, 313 pages. 1996.

Vol. 60: A. Kleusberg, P. J. G. Teunissen (Eds.), GPS for Geodesy. VII, 407 pages. 1996.

Vol. 61: M. Breunig, Integration of Spatial Information for Geo-Information Systems. XI, 171 pages. 1996.

Vol. 62: H. V. Lyatsky, Continental-Crust Structures on the Continental Margin of Western North America. XIX, 352 pages. 1996.

Vol. 63: B. H. Jacobsen, K. Mosegaard, P. Sibani (Eds.), Inverse Methods. XVI, 341 pages, 1996.

Vol. 64: A. Armanini, M. Michiue (Eds.), Recent Developments on Debris Flows. X, 226 pages. 1997.

Vol. 65: F. Sansò, R. Rummel (Eds.), Geodetic Boundary Value Problems in View of the One Centimeter Geoid. XIX, 592 pages. 1997.

Vol. 66: H. Wilhelm, W. Zürn, H.-G. Wenzel (Eds.), Tidal Phenomena. VII, 398 pages. 1997.

Vol. 67: S. L. Webb, Silicate Melts. VIII. 74 pages. 1997.

Vol. 68: P. Stille, G. Shields, Radiogenetic Isotope Geochemistry of Sedimentary and Aquatic Systems. XI, 217 pages. 1997.

Vol. 69: S. P. Singal (Ed.), Acoustic Remote Sensing Applications. XIII, 585 pages. 1997.

Vol. 70: R. H. Charlier, C. P. De Meyer, Coastal Erosion – Response and Management. XVI, 343 pages. 1998.

Vol. 71: T. M. Will, Phase Equilibria in Metamorphic Rocks. XIV, 315 pages. 1998.

Vol. 72: J. C. Wasserman, E. V. Silva-Filho, R. Villas-Boas (Eds.), Environmental Geochemistry in the Tropics. XIV, 305 pages. 1998.

Vol. 73: Z. Martinec, Boundary-Value Problems for Gravimetric Determination of a Precise Geoid. XII, 223 pages. 1998.

Vol. 74: M. Beniston, J. L. Innes (Eds.), The Impacts of Climate Variability on Forests. XIV, 329 pages. 1998.

Vol. 75: H. Westphal, Carbonate Platform Slopes – A Record of Changing Conditions. XI, 197 pages. 1998.

Vol. 76: J. Trappe, Phanerozoic Phosphorite Depositional Systems. XII, 316 pages. 1998.

Vol. 77: C. Goltz, Fractal and Chaotic Properties of Earthquakes. XIII, 178 pages. 1998.